'Impassioned ... engagingly written'
Saul David, *Daily Telegraph*

'Gladwell is to be applauded for insisting that we
look at this uncomfortable history a little more
honestly ... a very readable account. It is full of the
enlightening and entertaining tangents that have
become something of a Gladwell trademark'
Keith Lowe, *Spectator*

'Lively, engaging ... A fascinating story
appealingly delivered' Gerard DeGroot,
The Times

'Told with the muscular, driving narrative and fizzingly
charismatic (real-life) characters of a movie'
Ed Grenby, *Radio Times*

'Fascinating ... Gladwell's eloquence and flair
for lateral thinking make for a compelling read'
Simon Griffith, *Mail on Sunday*

'A thought-provoking, accessible account of how
people respond to difficult choices in difficult
times ... Gladwell's easy conversational style works
well ... his portraits of individuals are compelling'
Diana Preston, *Washington Post*

'Unexpected empathy ... fabulistic energy'
Esquire

'Important and characteristically readable ... Gladwell
is possibly the most confident storyteller in non-fiction.
He always knows exactly where he is going, and he
takes you with him in pleasure and comfort'
Simon Kuper, *New Statesman*

'Excellent revisionist history … another Gladwell everything-you-thought-you-knew-was-wrong page-turner'
Kirkus Reviews

'A ruminative, anecdotal account of what led up to the deadliest air raid of WWII … Gladwell provides plenty of colorful details and poses intriguing questions about the morality of warfare … fans will savor the insights into "how technology slips away from its intended path"'
Publishers Weekly

'Truly compelling … written in *New York Times* bestseller Malcolm Gladwell's characteristic approachable, story-telling style'
Zibby Owens, *Good Morning America*

'Malcolm Gladwell is a one-in-a-generation kind of writer … He has an uncanny way of finding the story within the story and pointing out the important lessons often hiding in plain sight'
Bryan Elliott, Inc.com

ABOUT THE AUTHOR

Malcolm Gladwell is the author of six international bestsellers: *The Tipping Point*, *Blink*, *Outliers*, *What the Dog Saw*, *David and Goliath* and, most recently, *Talking to Strangers*. He is the host of the podcast *Revisionist History*, a staff writer at *The New Yorker* and co-founder of the audio company Pushkin Industries. He graduated from the University of Toronto, Trinity College, with a degree in History. Gladwell was born in England and grew up in rural Ontario. He lives in New York.

The Bomber Mafia

A Tale of Innovation and Obsession

Malcolm Gladwell

PENGUIN BOOKS

PENGUIN BOOKS

UK | USA | Canada | Ireland | Australia
India | New Zealand | South Africa

Penguin Books is part of the Penguin Random House group of companies
whose addresses can be found at global.penguinrandomhouse.com

First published in the United States of America by Little, Brown and Company 2021
First published by Allen Lane 2021
Published in Penguin Books 2022
001

Printed and bound in Great Britain by Clays Ltd, Elcograf S.p.A.

The authorized representative in the EEA is Penguin Random House Ireland,
Morrison Chambers, 32 Nassau Street, Dublin D02 YH68

A CIP catalogue record for this book is available from the British Library

ISBN: 978-0-141-99837-4

www.greenpenguin.co.uk

Penguin Random House is committed to a
sustainable future for our business, our readers
and our planet. This book is made from Forest
Stewardship Council® certified paper.

To KMO (and BKMO!)

Contents

CONTENTS

Author's Note

As a little boy, lying in his bed, my father would hear the planes overhead. On their way in. Then, in the small hours of the morning, heading back to Germany. This was in England, in Kent, a few miles south and east of London. My father was born in 1934, which meant he was five when the Second World War broke out. Kent was called Bomb Alley by the British, because it was the English county that German warplanes would fly over on their way to London.

It was not uncommon, in those years, that if a bomber missed its target or had bombs left over, it would simply drop them anywhere on the return trip. One day, a stray bomb landed in my grandparents' back garden. It didn't explode. It just sat there, half buried in the ground—and I think it fair to say that if you were a five-year-old boy with an interest in things mechanical, a German bomb sitting unexploded in your backyard would have been just about the most extraordinary experience imaginable.

Not that my father described it that way. My dad was a mathematician. And an Englishman, which is to say that the language of emotion was not his *first* language. Rather, it was like Latin, or French—something one could study and understand but never fully master. No, that an unexploded German bomb in your backyard would be the most extraordinary experience imaginable for a five-year-old was *my* interpretation when my father told me the story of the bomb, when *I* was five years old.

That was in the late 1960s. We were living in England then, in Southampton. Reminders of what the country had gone through were still everywhere. If you went to London, you could still tell where the bombs had landed—wherever a hideous brutalist building had sprouted up on some centuries-old block.

BBC Radio was always on in our house, and in those days, it seemed like every second interview was with an old general or paratrooper or prisoner of war. The first short story I wrote as a kid was about the idea that Hitler was actually still alive and coming for England again. I sent it to my grandmother, the one in Kent who'd had the unexploded bomb in her back garden. When my mother heard about my story, she admonished me: someone who had lived through the war might not enjoy a plotline about Hitler's return.

My father once took me and my brothers to a beach overlooking the English Channel. We crawled together through the remnants of an old World War II fortification. I still remember the thrill of wondering whether we would come across some old bullets, or a shell casing, or even the skeleton of some long-lost German spy who'd washed up on shore.

I don't think we lose our childhood fascinations. I know I didn't. I always joke that if there's a novel with the word *spy* in it, I've read it. One day a few years back, I was looking at my bookshelves and realized—to my surprise—just how many nonfiction books about war I had accumulated. The big history bestsellers, but also the specialty histories. Out-of-print memoirs. Academic texts. And what aspect of war were most of those books about? Bombing. *Air Power,* by Stephen Budiansky. *Rhetoric and Reality in Air Warfare,* by Tami Davis Biddle. *Decision over Schweinfurt,* by Thomas M. Coffey. Whole shelves of these histories.*

Usually when I start accumulating books like that it's because I want to write something about the subject. I have shelves of books on social psychology because I've made my living writing about social psychology. But I never really wrote much about war—especially not the

---

* I could go on. If, for example, you haven't read Roberta Wohlstetter's *Pearl Harbor: Warning and Decision,* then you're missing a real treat.

Second World War or, more specifically, airpower. Just bits and pieces here and there.* Why? I don't know. I imagine that a Freudian would have fun with that question. But maybe the simpler answer is that the more a subject *matters* to you, the harder it is to find a story you want to tell about it. The bar is higher. Which brings us to *The Bomber Mafia*, the book you are reading now. I'm happy to say that with *The Bomber Mafia* I've found a story worthy of my obsession.

One last thing—about the use of that last word, *obsession*. This book was written in service to my obsessions. But it is also a story about other people's obsessions, about one of the grandest obsessions of the twentieth century. I realize, when I look at the things I've written about or explored over the years, that I'm drawn again and again to obsessives. I like them. I like the idea that someone could push away all the concerns and details that make up everyday life and just zero in on one thing—the thing that fits the contours of his or her imagination. Obsessives lead us astray sometimes. Can't see the bigger picture. Serve not just the world's but also their own narrow interests. But I don't think we get progress or innovation or joy or beauty without obsessives.

---

* Airpower has been something I've explored in a number of episodes of my podcast, *Revisionist History*, including "Saigon 1965," "The Prime Minister and the Prof," and the eponymous series starting with "The Bomber Mafia" in season 5.

When I was reporting this book, I had dinner with the then chief of staff of the US Air Force, David Goldfein. It was at the Air House, on the grounds of Joint Base Myer–Henderson Hall, in northern Virginia, just across the Potomac River from Washington, DC — a grand Victorian on a street of grand Victorians where many of the country's top military brass live. After dinner, General Goldfein invited a group of his friends and colleagues — other senior Air Force officials — to join us. We sat in the general's backyard, five of us in total. They were almost all former military pilots. Many of their fathers had been military pilots. They were the modern-day equivalents of the people you are going to read about in this book. As the evening wore on, I began to notice something.

Air House is just down the road from Reagan National Airport. And every ten minutes or so, a plane would take off over our heads. Nothing fancy: standard commercial passenger planes, flying to Chicago or Tampa or Charlotte. And every time one of those planes flew overhead, the general and his comrades would all glance upward, just to take a look. They couldn't help themselves. Obsessives. My kind of people.

THE BOMBER MAFIA

"*This isn't working. You're out.*"

1.

There was a time when the world's largest airport sat in the middle of the western Pacific, around 1,500 miles from the coast of Japan, on one of a cluster of small tropical islands known as the Marianas. Guam. Saipan. Tinian. The Marianas are the southern end of a largely submerged mountain range—the tips of volcanoes poking up through the deep ocean waters. For most of their history, the Marianas were too small to be of much interest or use to anyone in the wider world. Until the age of airpower, when all of a sudden they took on enormous importance.

The Marianas were in Japanese hands for most of the Second World War. But after a brutal campaign, they fell to the US military in the summer of 1944. Saipan was first, in July. Then Tinian and Guam, in August. When the Marines landed, the Seabees—the Navy's construction battalion—landed with them and set to work.

In just three months, an entire air base—Isely Field—was fully operational on Saipan. Then, on the island of Tinian, the largest airport in the world, North Field—8,500-foot runways, four of them. And following that, on Guam, what is now Andersen Air Force Base, the US Air Force's gateway to the Far East. Then came the planes.

Ronald Reagan narrated war films at the time, and one of those was devoted to the earliest missions of the B-29, known as the Superfortress. Reagan described the plane as one of the wonders of the world, a massive airship:

> With 2,200 horsepower in each of four engines. With a fuel capacity equal to that of a railroad tank car. A tail that climbed two stories into the air. A body longer than a Corvette. Designed to carry more destruction and carry it higher, faster, farther than any bomber ever built before. And to complete this mission, that's exactly what she was going to have to do.

The B-29 could fly faster and higher than any other bomber in the world and, more crucially, farther than any other bomber. And that extended range—combined with the capture of the Marianas—meant that for the first time since the war in the Pacific began, US Army Air Forces were within striking distance of Japan. A special unit was created to handle the fleet of bombers now parked in the Marianas: the Twenty-First Bomber Command, under the leadership of a brilliant young general named Haywood Hansell.

Throughout the fall and winter of 1944, Hansell launched attack after attack. Hundreds of B-29s skimmed over the Pacific waters, dropped their payloads on Japan, then turned back for the Marianas. As Hansell's airmen prepared to launch themselves at Tokyo, reporters and camera crews flew in from the mainland, recording the excitement for the folks back home.

Ronald Reagan again:

B-29s on Saipan were like artillery pointed at the heart of Japan…The Japs might just as well have tried to stop Niagara Falls. The Twenty-First Bomber Command was ready to hit its first target.

But then, on January 6, 1945, Hansell's commanding officer, General Lauris Norstad, arrived in the

Marianas. Things were still pretty primitive on Guam: headquarters were just a bunch of metal Quonset huts on a bluff overlooking the ocean. Both men would have been exhausted, not just from the privations of the moment but also from the weight of their responsibilities.

I once read a passage by the Royal Air Force general Arthur Harris about what it meant to be an air commander in the Second World War:

> I wonder if the frightful mental strain of commanding a large air force in war can ever be realized except by the very few who have experienced it. While a naval commander may at the most be required to conduct a major action once or twice in the whole course of the war, and an army commander is engaged in one battle say once in six months or, in exceptional circumstances, as often as once a month, the commander of a bomber force has to commit the whole of it every twenty-four hours…It is best to leave to the imagination what such a daily strain amounts to when continued over a period of years.

So there were Hansell and Norstad in Guam. Two war-weary airmen, facing what they hoped might be the war's final chapter. Hansell suggested a quick tour: Stand on the beach. Admire the brand-new runways, cut from the jungle. Chat about tactics, plans. Norstad said no. He had something more personal to discuss. And in a moment that would stay with Haywood Hansell for the rest of his life, Norstad turned to him: *This isn't working. You're out.*

"I thought the earth had fallen in—I was completely crushed." That's how, years later, Hansell described his feelings in that moment. Then Norstad delivered the second, deeper blow. He said, *I'm replacing you with Curtis LeMay.*

General Curtis Emerson LeMay, thirty-eight years of age, hero of the bombing campaigns over Germany. One of the most storied airmen of his generation. Hansell knew him well. They had served together in Europe. And Hansell understood immediately that this was not a standard leadership reshuffle. This was a rebuke, an about-face. An admission by Washington that everything Hansell had been doing was now considered wrong. Because Curtis LeMay was Haywood Hansell's antithesis.

Norstad offered that Hansell could stay on if he wished, to be LeMay's deputy, a notion Hansell considered so insulting that he could barely speak. Norstad told him he had ten days to finish up. Hansell walked around in a daze. On his last night in Guam, Hansell had a little more to drink than usual and sang for his men while a young colonel played the guitar: "Old pilots never die, never die, they just fly-y-y away-y-y-y."

7

When Curtis LeMay arrived for the changeover, he flew himself to the island in a B-29 bomber. "The Star Spangled Banner" was played. The airmen of the Twenty-First Bomber Command marched by for review. A public relations officer proposed a picture of the two of them to mark the moment. LeMay had a pipe in his mouth—he always had a pipe in his mouth—and didn't know what to do with it. He kept trying to put it in his pocket. "General," the aide said, "please let me hold your pipe while the picture is taken."

LeMay said, in a quiet voice, "Where do you want me to stand?" The cameras clicked and captured Hansell squinting off into the distance, LeMay looking down at the ground. Two men, anxious to be anywhere but in each other's company. And with that, it was over.

The Bomber Mafia is the story of that moment. What led up to it and what happened next—because that change of command reverberates to this day.

2.

There is something that has always puzzled me about technological revolutions. Some new idea or innovation comes along, and it is obvious to all that it will up-end our world. The internet. Social media. In previous generations, it was the telephone and the automo-bile. There's an expectation that because of this new invention, things will get better, more efficient, safer, richer, faster. Which they do, in some respects. But then things also, invariably, go sideways. At one moment, social media is being hailed as something that will allow ordinary citizens to upend tyranny. And then in the next moment, social media is feared as the platform that will allow citizens to tyrannize one another. The automobile was supposed to bring freedom and mobility, which it did for a while. But then millions of people found themselves living miles from their workplaces, trapped in endless traffic jams on epic commutes. How is it that, sometimes, for any number of unexpected and random reasons, technology slips away from its intended path?

The Bomber Mafia is a case study in how dreams go awry. And how, when some new, shiny idea drops down from the heavens, it does not land, softly, in our laps. It lands hard, on the ground, and shatters. The story I'm about to tell is not really a war story. Although it mostly takes place in wartime. It is the story of a Dutch genius and his homemade computer. A band of brothers in central Alabama. A British psychopath. Pyromaniacal chemists in a basement labs at Harvard. It's a story about the messiness of our intentions, because we always forget the mess when we look back.

And at the heart of it all are Haywood Hansell and Curtis LeMay, who squared off in the jungles of Guam.

Part One

The Dream

CHAPTER ONE

"Mr. Norden was content to pass his time in the shop."

1.

Back when the war that would consume the world was a worry but not yet a fact, a remarkable man came to the attention of the US military.

His name was Carl L. Norden. Throughout his life, Norden shunned the limelight. He worked alone— sometimes returning to Europe during crucial periods to tinker and dream at his mother's kitchen table. He built a business with hundreds of employees. Then when the war was over, he left it all behind. There are no full-length biographies of Norden. No profile pieces.[*]

---

[*] In 2011 I gave a TED Talk on Norden and his invention.

No statues in his honor. Not in his native Holland; not in Switzerland, where he lived out his days; and not in downtown Manhattan, where he did his most important work. Norden influenced the course of a war and sparked a dream that would last the remainder of the century. It does not seem possible that someone could have left as much of a mark on his world as Norden did and then disappear from sight. Yet he did. In one 352-page technical book about Norden's invention, there is a single sentence devoted to him: "Mr. Norden was content to pass his time in the shop, which sometimes was an eighteen-hour day."

That's it.

So before we start in on Norden's dream and its consequences—the effect Norden would have on an entire generation—let us start with Norden himself. I asked Professor Stephen L. McFarland, one of the few historians—maybe the only historian—who has really dug into the story of Carl Norden, why there's so little documentary record about the inventor. The professor replied that it is "primarily because he demanded absolute secrecy." He went on to describe the man: "Well, he was extremely prickly. His ego was greater than [that of] any person I've never met. And I said 'never met' because of course I never met Norden."

Norden was Dutch. He was born in what is now Indonesia, then a Dutch colony. He spent three years apprenticing in a Swiss machine shop, then got an engineering degree from Zurich's prestigious Federal Polytechnic School, where one of his classmates was Vladimir Lenin. Norden was trim, dapper. He wore a three-piece suit. Had short white hair with a little cowlick, a thriving mustache, and heavy-lidded eyes underwritten with deep lines, as if he hadn't slept in years. His nickname was Old Man Dynamite. He drank coffee by the gallon. Lived on steak.

As McFarland explained,

> He truly believed in a very biological sense that sun created stupidity. And so you would never see him outside without a big hat on. His family always was forced to wear hats outside. He was, as a young boy, stationed in the Dutch East Indies, and yet he and his family always wore hats because the sun caused stupidity.

McFarland wrote that Norden "read Dickens avidly for revelations on the lives of the disadvantaged and Thoreau for the discussion of the simple life." He hated paying taxes. He thought Franklin Roosevelt was the devil.

McFarland described how cranky Norden could be:

> There's a famous story where he was looking over a technician's shoulder and the technician got a little bit nervous and tried to strike up a

conversation, looking at him and saying, "Perhaps you could explain why we're making this part this way." And Norden screamed at the top of his lungs at him, after he yanked the cigar out of his mouth, and said, "There's a hundred thousand reasons why I designed that part that way. And none of it is your damn business." So that's how he treated all his employees. He was truly an Old Man Dynamite.

McFarland went on to explain Norden's perfectionism:

Expense didn't matter—it was "Make it as perfect as possible." I'd seen how engineers know what they know and how they do what they do, but all of them talked about the importance of studying what had been done before. Norden's attitude was, "I don't want to hear about it." All he wanted was blank sheets of paper, a pencil, and a couple of engineering books that were filled with formulas about how to calculate certain mathematical problems. He was a true believer in blank slate, and this reveals his ego. He said, "I don't want to know the mistakes other people made. I don't want to know what they did right. I'm going to develop what's right myself."

What was Carl Norden developing on his blank sheets of paper? A bombsight. A bombsight is not something that anyone uses anymore—not in the age of radar and GPS—but for the better part of the last century, bombsights were matters of great importance. Let me go further, because there is a real risk here of understatement. If you were to have made a list in, say, the early years of the twentieth century of the ten biggest unsolved technological problems of the next half century, what would have been on that list? Well, some things are obvious. Vaccines were desperately needed to prevent childhood diseases—measles, mumps. Better agricultural fertilizers were needed to help prevent famine. Huge parts of the world could be made more productive with affordable, convenient air-conditioning. A car cheap enough for a working-class family to afford. I could go on. But somewhere on that list would be a military question—namely, is there a more accurate way to drop a bomb from an airplane?

Now, why does that problem belong on the same list as vaccines and fertilizers and air-conditioning? Because early in the twentieth century, the world went through World War I, in which thirty-seven million people were wounded or killed. Thirty-seven million. There were over a million casualties in the Battle of the Somme, a single battle that had no discernible point or impact on the course of the war. For those who lived through it, World War I was a *deeply* traumatic experience.

So what could be done? A small group of people came to believe that the only realistic solution was for armies to change the way they fought wars. To learn to fight—if this doesn't sound like too much of an oxymoron—*better* wars. And the people who made the argument for better wars were pilots. Airmen. People obsessed with one of the newest and most exciting technological achievements of that era—the airplane.

2.

Airplanes made their first big appearance in World War I. I'm sure you've seen pictures of those early planes. Plywood, fabric, metal, and rubber. Two wings, upper and lower, connected by struts. One seat. A machine gun facing forward, synchronized to fire through the propeller. They resembled something that came in the mail to be assembled in a garage. The most famous of World War I fighter planes was the Sopwith Camel. (That's the one that Snoopy flew in the old *Peanuts* comic strip.) It was a mess. "In the hands of a novice," the aviation writer Robert Jackson says, "it displayed vicious characteristics that could make it a killer." Meaning a killer of the pilot flying it, not the enemy under attack. But a new generation of pilots looked at these contraptions and said, *Something like this can make all that deadly, wasteful, pointless conflict on the ground obsolete. What if we just fought wars from the air?*

One of those airmen was a man named Donald Wilson. He served in the First World War and remembered the fear that had gripped his fellow soldiers.

As he recounted in an oral history in 1975:

One fellow killed himself and chose our mess hall as the place to do it. Put his mouth over the muzzle of his rifle and pulled the trigger. And another man while we were in the trenches shot himself in the leg. So those people must have magnified their ideas of the great danger. But I think by and large, the most of us just didn't realize what we were getting into.

Wilson started flying in the 1920s and ended up as a general in the Second World War. I ran across a memoir that Wilson self-published in the 1970s. It's called *Wooing Peponi,* and it looks like a high school yearbook. It goes on forever. And right in the middle, Wilson has this strangely riveting passage about the conclusion he came to in his first years of flying: "Then out of nowhere a vision evolved. As in later years, in entirely different context, Martin Luther King said, in a moving speech, 'I had a dream.'"

Wilson is comparing his vision of the promise of airpower to the most iconic moment in the civil rights movement. And then he borrows King's rhetorical pattern as well:

> I had a dream…that nations fought each other in order to dictate terms and not to prove supremacy of arms, as military tradition insisted. I had a dream that important nations, the likely adversaries, were industrialized and dependent upon smooth operation of organized and mutually sustaining elements. I had a dream that the new and coming air capability could destroy a limited number of targets within this web of interdependent features of the modern nation. I had a dream that such destruction and the possibility of more of the same, would cause the victim to sue for peace.

In every way, this passage is audacious. There were so few military pilots in the United States back in those days that they all knew each other. It was like a club. A band of zealots. And Wilson said this tiny club with its ramshackle flying machines could reinvent war.

"I had a dream that such destruction and the possibility of more of the same, would cause the victim to sue for peace"? That means he believed that planes could win wars all by themselves. They could swoop down and bomb select targets and bring the enemy to its knees without the slaughter of millions on a battlefield.

But before the dream could be made real, the airmen knew they had to deal with a problem, a very specific technical problem, a problem so consequential that it belongs on the top ten list of problems, along with vaccines and fertilizers. If you thought, as the dreamers did, that the airplane could revolutionize warfare—could swoop down and hit select targets and bring the enemy to its knees—then you had to have a way to hit those select targets from the air. And no one knew how to do that.

I asked Stephen McFarland why it is so difficult to pinpoint a bombing target. His response:

> It's amazing to me. I mean, I just assumed that you've watched the videos and the movies. And they say, "Just put the crosshairs on the target, and the bombsight will do the rest." But there's an amazing number of elements that [go] into dropping a bomb accurately on a target. If you think about your own car, driving down the highway at sixty, seventy miles an hour, you can imagine throwing something out the window and trying to hit something, even if it's stationary like a sign or a tree or anything on the side of the road. You get an idea of just how hard that is.

21

If you're trying to throw a bottle into a garbage can from a car going fifty miles per hour, you have to perform some physics calculations on the fly: the garbage can is stationary, but you and the car are moving quickly, so you have to release the bottle well before you reach the can. Right? But if you're in an airplane at twenty thousand or thirty thousand feet, the problem is infinitely more complicated.

McFarland went on:

Aircraft in World War II were flying at two hundred, three hundred miles an hour, sometimes as fast as five hundred miles an hour. They were dropping bombs from up to thirty thousand feet. That would take between twenty and thirty, [maybe] thirty-five seconds to hit the ground. And during that whole time, you're being shot at. You're having to look through clouds or ... [avoid] antiaircraft artillery. You're having to deal with factory decoys, smoke screens. There's the smoke from other bombs, people screaming in your ear, the excitement, all these strange things that happen once war begins.

The wind could be blowing at a hundred miles per hour. You'd have to factor that in. If it's cold, the air is dense, and the bomb will fall slowly. If it's warm, the air is thin, and the bomb will fall fast. Then you'd also have to consider: Is the plane level? Is it moving from side to side? Or up and down? A tiny degree of error at the release point could translate to a big error on the ground. And from twenty thousand feet, can you even see the target? A factory might be big and obvious up close, but from that far up, it looks like a postage stamp. Bombers, in the early days of aviation, couldn't hit anything. Not even close. The bombardiers might as well have been throwing darts at a dartboard with their eyes closed. The dream that the airplane could revolutionize warfare was based on a massive untested and unproven assumption: that somehow, someone at some point would figure out how to aim a bomb from high in the sky with something close to accuracy. It was a question on the era's technological wish list. Until...Carl Norden.

McFarland says Norden's design methods were singular:

> He had no help. He did it all by himself. It was all in his mind. He didn't carry notes. He didn't have a notepad. You can't go to his archive. There is no such thing. It was all kept in his head, and for a man to keep that kind of complexity in his head...I was just amazed that it could be done that way. But engineers refer to something called "the mind's eye," that they see things in their mind, not with their eyes, but with their mind's eye. And that was truly Carl Norden.

I asked McFarland if he thought Norden was a genius. His reply:

> Well, he would tell you that only God invents; humans discover. So for him, it was not "genius." He would have refused to accept that term. He would not appreciate it, would not accept anyone calling him a genius. He would say he's just one who discovers the greatness of God, the creations of God; that God reveals truths through people who are willing to work hard and to use their minds to discover God's truths.

Norden began working on the bombsight problem in the 1920s. He got a Navy contract—although he would later work for the Army Air Corps, which is what the US Air Force was called in those days. He set up shop on Lafayette Street in the part of Manhattan now called SoHo. And there he began work on his masterpiece.

By the time the United States entered World War II, the military rushed to equip its bombers with the Norden bombsight. Those bombers, in most cases, had a crew of ten men: pilot, copilot, navigator, gunners, and, most crucially, bombardiers, the people who aimed and dropped the bombs. If the bombardier did not do his job, then the efforts of all nine of his crewmen were wasted.

A wartime military training film for bombardiers explained the importance of the Norden bombsight by showing aerial photographs of enemy targets:

> One of them may be your target. They are the reason for your being here. The reason for all the vast equipment assembled in this and other bombardier schools. For the instructors here to train you. For the pilots here to fly you on your missions.
>
> In all likelihood some one of you now sitting in this room will see one of these targets, not projected on a screen but moving under the crosshair of your bombsight. And where will they fall, those bombs of yours?...One hundred feet off? Five hundred feet? That will depend on how well you'll have taught your fingers and your eyes to match the precision that has been built into your Norden bombsight.

Its official name was the Mark XV. It was dubbed "the football" by the airmen who used it. It weighed fifty-five pounds. It sat on a kind of platform—a packing box, stabilized by a gyroscope—that kept it level at all times, even as the plane was bouncing around. The bombsight was essentially an analog computer, a compact, finely machined contraption composed of mirrors, a telescope, ball bearings, levels, and dials. From a moving plane, the bombardier peered through the telescope at the target and made a fantastically complicated series of adjustments. Norden created sixty-four algorithms that he believed addressed every question of the bombing problem, including: How much do the speed and direction of the wind affect the trajectory of a bomb? How much does the air temperature affect it? Or the speed of the aircraft? To be properly trained on the Norden took six months.

Just watching the Army training film is enough to hurt your head. The narrator says,

Now look at the line in the flooring. That was your sighting line when you started. Goes straight to the target. I know: when you're up in the air there aren't any nice convenient lines drawn in the ground to help you. Your bombsight, though, gives you the equivalent of them. Remember how the sight's made in two parts? Underneath, there's the stabilizer. And in that there's another gyro, only it has a horizontal axis.

Above that is your sight. The stabilizer is fixed in the longitudinal axis of the airplane. But you can keep turning the sight so that it's always pointing at the target. But the sight is also connected to the stabilizer by rods. By these, the gyro controls the position of the sight, so that no matter how much the airplane yaws, the sight will always point in the same direction.

All this so the bombardier could know exactly when to shout, "Bombs away!"

McFarland explained one of the fine points of Norden's work:

One of Norden's sixty-four algorithms compensated for the fact that when you drop a bomb it takes thirty seconds to hit the target. During those thirty seconds, the earth actually moves as it spins on its axis.

So he actually created a formula. If it was going to take twenty seconds for the bomb to hit the target, then the earth would move—I'm going to make up a number—twelve feet. You therefore had to adjust the computer to [the fact that] the target's now moved twelve feet. If you're at twenty thousand feet, it might move twenty-five feet. And all of these then had to be put into this computer.

The Army bought thousands of Norden bombsights. Before every mission, the bombardier, with an armed escort, would retrieve his device from a vault. He would carry it out to the plane in a metal box. In the event of a crash landing, the bombardier was instructed to destroy the bombsight immediately, lest it fall into enemy hands. Legend has it that bombardiers were even given an eighteen-inch-long explosive device to do the trick. And, as a final precaution, they had to take a special oath: "I solemnly swear that I will keep inviolate the secrecy of any and all confidential information revealed to me, and in full knowledge that I am a guardian of one of my country's most priceless assets, do further swear to protect the secrecy of the American bombsight, if need be, with my life itself."

In the middle of all this drama and secrecy was Carl Norden. Maddening, eccentric Norden. Before the United States entered the war, while he was still perfecting his invention, he would sometimes leave Manhattan and return to his mother's house, in Zurich. McFarland said this would put US officials "up in arms":

The FBI sent agents with him to try to protect him. The British supposedly thought that he was working for the Germans as a spy. And [the Army was] afraid that the British would try to capture him. But he absolutely insisted. He said, *I'm going to Switzerland. There's nothing you can do to stop me.* And of course, the laws of wartime were not yet in effect because the United States wasn't in the war. So legally there was no way they could stop him.

Why did the military put up with him? Because the Norden bombsight was the Holy Grail.

Norden had a business partner named Ted Barth.

He was the salesman, the public face. And he claimed, the year before the United States joined the war, that "We do not regard a fifteen-foot square...as being a very difficult target to hit from an altitude of thirty thousand feet." The shorthand version of that—which would serve as the foundation of the Norden legend— was that the bombsight could drop a bomb into a pickle barrel from six miles up.

To the first generation of military pilots, that claim was intoxicating. The most expensive single undertaking of the Second World War was the B-29 Bomber, the Superfortress. The second most expensive was the Manhattan Project, the massive, unprecedented effort to invent and build the world's first atomic bomb. But the third most expensive project of the war? Not a bomb, not a plane, not a tank, not a gun, not a ship. It was the Norden bombsight, the fifty-five-pound analog computer conceived inside the exacting imagination of Carl L. Norden. And why spend so much on a bombsight? Because the Norden represented a dream—one of the most powerful dreams in the history of warfare: if we could drop bombs into pickle barrels from thirty thousand feet, we wouldn't need armies anymore. We wouldn't need to leave young men dead on battlefields or lay waste to entire cities. We could reinvent war. Make it precise and quick and almost bloodless. Almost.

CHAPTER TWO

"We make progress unhindered by custom."

1.

Revolutions are invariably group activities. That's why Carl Norden was such an anomaly. Rarely does someone start a revolution alone, at his mother's kitchen table. The impressionist movement didn't begin because one genius took up painting impressionistically and, like the Pied Piper, attracted a trail of followers. Instead, Pissarro and Degas enrolled in the École des Beaux-Arts at the same time; then, Pissarro met Monet and, later, Cézanne at the Académie Suisse; Manet met Degas at the Louvre; Monet befriended Renoir at Charles Gleyre's studio; and Renoir, in turn, met Pissarro and Cézanne; and soon enough everyone was hanging out at the Café Guerbois, trading ideas and egging each other on, and sharing and competing and dreaming, all together, until something radical and entirely new emerged.

This happens all the time. Gloria Steinem was the most famous face of the feminist movement in the early 1970s. But what was it that led to a doubling of the number of women elected to office in the United States? Gloria Steinem plus Shirley Chisholm, Bella Abzug, and Tanya Melich coming together to create the National Women's Political Caucus. Revolutions are birthed in conversation, argument, validation, proximity, and the look in your listener's eye that tells you you're on to something.

For those caught up in the dream of changing modern warfare, that place where friends spent time with one another and had long arguments into the night and saw that look in their comrades' eyes was an air base called Maxwell Field. Maxwell Field was—and is—in Montgomery, Alabama. It was an old cotton plantation converted to an airfield by the Wright brothers, Orville and Wilbur. In the 1930s it became home to something called the Air Corps Tactical School, the aviation version of the Army War College in Carlisle, Pennsylvania, or the Naval War College in Newport, Rhode Island. Much of the base today remains the same as it was when it was built, in the 1930s: everything is in pale yellow concrete or stucco, with red tile roofs. There are hundreds of elegant houses for the officers, built in the French provincial style on quiet curving streets lined with giant ring-cup oak trees. In the summer, the air is thick and wet. This is deep inside Alabama. The grand nineteenth-century buildings that make up the Alabama state legislature are just down the road, a few miles away. It does not feel like the birthplace of a revolution.

But it was.

In those years, the Air Force was not a separate branch of the military. It was a combat division of the Army. It existed to serve the interests of the ground forces. To support, assist, accompany. The legendary Army general John "Black Jack" Pershing, who commanded the American forces in World War I, once wrote of airpower that it "can of its own account neither win a war at the present time nor, so far as we can tell, at any time in the future."[*] That's what the military establishment thought of airplanes. Richard Kohn, chief historian of the US Air Force for

---

[*] This quotation comes from a 1920 letter Pershing sent to the director of the Air Service, in which he argued that the Air Service should "remain a part of the Army." He believed that air forces existed to aid the Army and should remain under its command: "If success is to be expected, the military air force must be controlled in the same way, understand the same discipline, and act in accordance with the Army command under precisely the same conditions as other combat arms."

a decade, explains that in the early days, people just didn't understand airpower:

> I remember one congressman being quoted as saying, "Why do we have all this controversy over airplanes? Why don't we just buy one of them and let the services share it?"

The very first site of the Air Corps Tactical School was not in Alabama but in Langley, Virginia. There were stables out by the airplane hangars, and pilots were expected to learn how to ride, as if it were still the nineteenth century. Can you imagine how the Army's pilots of that era—and there were only a few hundred of them—felt about that? So long as they were part of the Army, they came to believe, they would be under the command of people who couldn't fly airplanes, didn't understand airplanes, and wanted them to rub down the horses every morning. The pilots wanted to be independent. And the first step toward independence was to move their training school as far away from the influence of the Army—culturally and physically— as humanly possible. The fact that Maxwell Field was on an old cotton plantation in a sleepy corner of the South was, to use the modern expression, a feature and not a bug.

Because airpower was young, the faculty of the Tactical School was young—in their twenties and thirties, full of the ambition of youth. They got drunk on the weekends, flew warplanes for fun, and raced each other in their cars. Their motto was: *Proficimus more irretenti*: "We make progress unhindered by custom." The leaders of the Air Corps Tactical School were labeled "the Bomber Mafia." It was not intended as a compliment—these were the days of Al Capone and Lucky Luciano and shoot-outs on the streets. But the Air Corps faculty thought the outcast label quite suited them. And it stuck.

Harold George, one of the spiritual leaders of the Bomber Mafia, put it like this: "We were highly enthusiastic; we were starting on, like, a crusade…knowing that there were a dozen of us and the only opposition we had was ten thousand officers and the rest of the Army, rest of the Navy."

George was from Boston. He joined the Army during the First World War and became captivated by airplanes. He started teaching at the Tactical School in the early 1930s and rose to the rank of general during the Second World War. After the war, he went to work for Howard Hughes, setting up Hughes's electronics business. Then George left to help build another electronics firm that became a giant defense contractor. And this is my favorite part: he was twice elected mayor of Beverly Hills.

That's one man. In one lifetime. But if you had asked Harold George what was the highlight of his career, he probably would have said those heady days in the 1930s, teaching at Maxwell Field.

As he said in an oral history in 1970, "Nobody seemed to understand what we were doing, and therefore we got no directives that we were to stop the kind of instruction that we were giving."

The Tactical School was a university. An academy. But not many of the faculty had any experience teaching. And the things they were teaching were so new and radical that there weren't really any textbooks for anyone to study or articles for anyone to read. So they mostly made things up—on the fly, so to speak. Lectures quickly turned into seminars, which turned into open discussions, which spilled out into dinner in the evening. That's what always happens: Conversation starts to seed a revolution. The group starts to wander off in directions in which no one individual could ever have conceived of going all by himself or herself.

Donald Wilson was another of the Bomber Mafia inner circle. He was the one who later wrote in his memoirs that he had a dream of a different kind of war. As he recalled of those days,

I feel quite certain that if the controlling element of the War Department general staff had known what we were doing at Maxwell Field, we would have all been put in jail. Because it was just so contrary to their established doctrine that I just can't imagine their knowing and allowing us to do it.

2.

When people thought about military aircraft in the first part of the twentieth century, they thought of fighter planes: small and highly maneuverable airplanes that could engage the enemy in the air. But not the renegades at Maxwell Field. They were obsessed with the technological advances in aviation that happened during the 1930s. Aluminum and steel replaced plywood. Engines got more powerful. Planes got bigger and easier to fly. They had retractable landing gear and pressurized fuselages. And those advances allowed the Bomber Mafia to imagine an entirely new class of airplane—something as large as the commercial airliners that had just started ferrying passengers across the United States. A plane that big and powerful wouldn't be limited to fighting other planes in the sky. It could carry bombs: heavy, powerful explosives that could do significant damage to the enemy's positions on the ground.

Now, why would that be so devastating? Because if you put one of these newly powerful engines inside one of these newly massive airplanes, that plane could fly so far and so fast for so long that nothing could stop it. Antiaircraft guns would be like peashooters. Enemy fighters would be like small annoying gnats, buzzing harmlessly. This kind of airplane could have armor plating, guns at the back and front to defend itself. And so we arrive at principle number one of the Bomber Mafia doctrine: The bomber will always get through.

The second tenet: Up until then, it had been assumed that the only way to bomb your enemy was in the safety of darkness. But if the bomber was unstoppable, why would stealth matter? The Bomber Mafia wanted to attack by daylight.

The third tenet: If you could bomb by daylight, then you could *see* whatever it was you were trying to hit. You weren't blind anymore. And if you could see, it meant that you could use a bombsight—line up the target, enter the necessary variables, let the device do its work—and *boom*.

The fourth and final tenet: Conventional wisdom said that when a bomber approached its target, it had to come down as close as it could to the ground in order to aim properly. But if you had the bombsight, you could drop your bomb from way up high—outside the range of antiaircraft guns. *We can drop a bomb into a pickle barrel from thirty thousand feet.*

High altitude. Daylight. Precision bombing. That was what the Bomber Mafia cooked up in its hideaway in central Alabama.

Historian Richard Kohn described the Bomber Mafia this way:

It was collegial. I would call it almost to the point of "band of brothers." But if you didn't buy the doctrine, and some of them didn't, you could be…not exactly expelled from the brotherhood but suspected and opposed.

There was a pilot on the Tactical School staff named Claire Chennault, who dared to challenge the Bomber Mafia orthodoxy. They ran him out of town.

Kohn continued: "They were a rebellious bunch. They engaged in public relations campaigns. Some of them wrote under pseudonyms to promote airpower."

I didn't really grasp the audacity of the Bomber Mafia's vision until I went to Maxwell. It's now Maxwell Air Force Base, not Maxwell Field. It's home to Air University, the successor to the Air Corps Tactical School. People come from around the world to study there. The faculty includes many of the country's leading military historians, tacticians, and strategists. And I sat one afternoon with a group of Maxwell faculty in a conference room just a stone's throw from the place where the Bomber Mafia held forth almost a century ago. All the records from the original Tactical School are in the Maxwell archives, and the historians I spoke to had been through the Bomber Mafia's old field notes and lectures. They spoke of Donald Wilson and Harold George as if they were contemporaries. They *knew* them. I was struck, though, by one difference. A number of the historians I met with were themselves former Air Force pilots. They'd flown advanced fighter jets and stealth bombers and multimillion-dollar transport planes, so when they talked about airpower, they were talking about something tangible, something they had personal experience with.

But back in the 1930s, the Bomber Mafia was talking about something theoretical, something they *hoped* would exist.

It was a dream.

Richard Muller, professor of airpower history at Air University, put it like this:

> There's nothing on the ramp that can match what they're thinking. They're on crack cocaine. You can kind of ask yourself if you go to a museum, an aviation museum—go down to Pensacola or go to the [National] Air and Space Museum or Wright-Patt[erson Air Force Base] and look at the planes that are on the field in the early thirties, when this idea first comes, and you go, What the hell? How much cocaine are those guys snorting?

One of the unexpected pleasures of talking to military historians is their irreverence toward their own institutions. Muller continued:

There was just this faith that they'll get there. They don't quite know how. They don't quite know where, but they'll get there, and it's not particularly unreasonable in their own time and place. It's not unreasonable for them to have this kind of faith. But really one of the central things that happens inside of this group is a belief in technological progress and material development, and that they can get the right plane. They go from the B-9 to the B-10 to the B-12 to the B-15 prototype to the B-17 to the B-29 in about ten years, which is extraordinary when you think about it.

3.

I worry that I haven't fully explained just how radical— how revolutionary—the Bomber Mafia thinking was. So allow me a digression. It's from a book I've always loved called *The Masks of War*, by a political scientist named Carl Builder. Builder worked for the RAND Corporation, the Santa Monica–based think tank set up after the Second World War to serve as the Pentagon's external research arm.

Builder argued that you cannot understand how the three main branches of the American military behave and make decisions unless you understand how *different* their cultures are. And to prove this point, Builder said, just look at the chapels on each of the service academy campuses.

The chapel at West Point military academy, the historic training ground for the officers of the US Army, stands on a bluff high above the Hudson River, dominating the skyline of the campus. The chapel was completed in 1910, in the grand Gothic revival style. It is built entirely out of somber gray granite, with tall, narrow windows. It has the brooding power of a medieval fortress—solid, plain, unmovable. Builder writes, "This is a quiet place for simple ceremonies with people who are close to each other and to the land that has brought them up."

That's the Army: deeply patriotic, rooted in service to country.

Then there's the chapel at the Naval Academy, in Annapolis. It was built almost at the same time as its West Point counterpart, but it's much bigger. Grander. It's in the style of American Beaux-Arts, with a massive dome based on the design of the military chapel at Les Invalides, in Paris. The stained-glass windows are enormous, letting the light shine into the ornate, detailed interior. That's very Navy: arrogant, independent, secure in the global scale of its ambitions.

Compare those two to the cadet chapel at the Air Force Academy, in Colorado Springs. This is a chapel from another universe. It was finished in 1962, but if I told you that it was finished last month, you would say, "Wow, that's a futuristic building." The Air Force chapel looks like someone lined up a squadron of fighter jets like dominoes with their noses pointed toward the heavens. It looks ready to take flight with a magnificent, deafening *whoosh*. Inside the cathedral, there are more than twenty-four thousand pieces of stained glass, in twenty-four different colors, and at the front, a cross forty-six feet tall and twelve feet wide, with crossbeams that look like propellers. Outside, four fighter jets are jauntily parked, as if some pilots, on a whim, had dropped by for Sunday morning communion.

The chapel's architect was a brilliant modernist out of Chicago named Walter Netsch. He was given the same creative freedom and limitless budget that the Air Force usually gives to the people who come up with stealth fighters.

In a 1995 interview, Netsch recalled the commission:

I came home with this tremendous feeling of: How can I in this modern age of technology create something good to be as inspiring and aspiring as Chartres…? In the meantime, I had gotten this idea here in Chicago, working with my engineer, of the tetrahedrons and compiling the tetrahedrons together.

What do you think it says about the Air Force that they would construct a cathedral out of aluminum and steel, in the shape of an upright fighter jet, in the middle of the Colorado mesa? That's what Carl Builder asked in his book. And his conclusion was: this is a group of people who desperately want to differentiate themselves as much as possible from the older branches of military service, the Army and the Navy. And, further, the Air Force is utterly uninterested in heritage and tradition. On the contrary, it wants to be modern.

Netsch designed the entire Air Force Academy chapel around pyramid-shaped seven-foot modules. Tetrahedrons! This is a branch of the service for people who want to start over, to wage war in new ways, to ready themselves for today's battles. They aren't spending their time studying the Peloponnesian War or the Battle of Trafalgar. The Air Force is obsessed with tomorrow, and with how technology will prepare it for tomorrow. And what happens with Netsch's chapel after it's built? It has all kinds of structural problems. Of course it does! Like some brilliant bit of breakthrough computer code, it had to be debugged.

Netsch explained:

You get into technology, you sometimes get into trouble...What happened is that all of a sudden, these leaks started. And [we] would fly out to Colorado Springs and check in [to] a little cheap motel and wait for the rains. And it would rain, and we would rush up to the chapel—it's a big building—and try to find out where it was leaking inside…I had to write a report, and I was so hurt about these leaks. I called it "A Report on Water Migration on the Air Force Academy Chapel." Needless to say, I received humorous digs over my euphemism. But what we found out was that…each of the tetrahedral groups would move in the wind. It's very windy up there, and the building can receive wind from many planes. And it's long, so it could be doing one thing at one end and another thing at the other end. These joints where everything is connected is where all the glass goes through.

So it was finally decided that what we should do is develop a big cover of plastic over the glass windows, which eliminated many of the sources of the problem, because each little piece of glass sitting in that window frame, everything begins to—it doesn't take much for water to come through. And so they went and put in these long panels of plastic, and it's done a lot to eliminate the major problem.

This is *so* Air Force. You build a twenty-first-century chapel in the middle of the twentieth century, and it's so far ahead of its time that you have to do an engineering workaround based on a reanalysis of mete-orological patterns. My point is—where did this radical new mind-set come from? It came from the Air Corps Tactical School, in that intellectual flurry between 1931 and 1941. In those seminar rooms and late-night arguments, the culture of the modern Air Force was born. They would take warfare into the air. They would leave every other branch of the service behind. And if you stand in the sanctuary of the Air Force Academy chapel and stare up at the soaring aluminum ribs of the ceiling, you'll get it.

Meanwhile, what's happening back at the Naval Academy? They're burnishing the brass rails in their chapel by hand.

4.

As with all revolutionary groups, the Bomber Mafia has a defining legend, an origin story. And as with all legends, it may not be strictly accurate, but here's how it goes:

On St. Patrick's Day in 1936, in Pittsburgh, there was a flood. It was a devastating event. Pittsburgh is unusual in that it sits at the head of a major river, the Ohio, formed by the convergence of two other rivers, the Monongahela and the Allegheny. And that day, the convergence of the rivers swelled in a massive flood.

Airmen do not typically concern themselves with land-based disasters. Hurricanes, maybe. Thunderstorms. A flood is the kind of thing the Army worries about. But there was an odd consequence of the Pittsburgh flood that would end up having a dramatic influence on the revolution brewing down at Maxwell Field. It had to do with the fact that among the hundreds of buildings along the riverbanks destroyed by the rising water was a factory belonging to a firm named Hamilton Standard. Hamilton Standard was the country's principal manufacturer of a spring used in making variable-pitch propellers, which were basic equipment on most airplanes at the time. But because Hamilton Standard couldn't make variable-pitch propeller springs, no one could make variable-pitch propellers, and because no one could make variable-pitch propellers, no one could make airplanes. The Pittsburgh flood brought the whole aeronautics industry of 1936 to a halt: for the want of a spring, the airplane business was lost.

Down in Alabama, the Bomber Mafia looked at what happened to Hamilton Standard, and the men's eyes lit up. The member of the Bomber Mafia who spent the most time thinking about that spring factory was Donald Wilson. And what happened in Pittsburgh made him realize something. War, in its classical definition, is the application of the full weight of military forces against the enemy until the enemy's political leadership surrenders. But Wilson thought—is that really necessary? If we just take out the propeller-spring factory in Pittsburgh, we cripple their air force. And if we can find another dozen or so crucial targets just like that—"choke points" was the phrase he used—bombing could cripple the whole country. Wilson then devised one of the Bomber Mafia's most famous thought experiments. And remember, the men could only *do* thought experiments. They didn't have any real bombers. Or any real enemy. Or any real resources. They were spitballing.

In the thought experiment, Wilson made the manufacturing hub of the Northeast the target:

Now, when we began theorizing about this thing...we had no air intelligence of any possible enemy. Thereby we had a thing...a unit that possibly could be reached by an enemy. And in order to illustrate this concept, we assume that an enemy would plant himself down in Canada and be within reach of this northeastern industrial area.

So the enemy in this thought experiment is in Canada—let's say Toronto. Toronto is 340 miles from New York City as the crow flies, easily within the range of the planes the Bomber Mafia was dreaming about. What kind of damage could a fleet of bombers do, coming down from Toronto on a single bombing run?

In a two-day presentation in April of 1939 at the Tactical School, they tried to figure it out.

I spoke about the thought experiment with historian Robert Pape, who wrote a book called *Bombing to Win*, about the origins of many of the ideas taught at the Air Corps Tactical School. Pape described the presentation:

> The bombing that they're focusing on [is], number one, the bridges. Number two, they have the bombing of the aqueducts. The bombing of the aqueducts is important because what they want to do is cause massive thirst in the New York population. They basically want to create a situation where there's almost no potable water for the population to drink. And then, number three, they target electric power.
>
> They're not really investigating the psychology of bombing. They're not investigating the sociology of bombing. They're not really even investigating the politics of the bombing—that is, the implications the bombing would have for populations, societies, and for governments. What they're really doing is focusing on the technology of the bombing of the time, what target sets it would allow the bombers to hit.

The presentation was given by a key Bomber Mafia associate, Muir Fairchild. Fairchild argued that the aqueducts are the most obvious targets. The aqueduct system serving New York City is ninety-two miles long. Then there's the power grid. Fairchild directed his students to a chart: "The Aerial Bomb Versus Traction Electric Power in the New York City Area."

As Fairchild concluded: "We see then that seventeen bombs, if dropped on the right spots, will not only take out practically all of the electric power of the entire metropolitan area but will prevent the distribution of outside power!"

Seventeen bombs! Conventional wisdom was that you would have to bomb the whole city—reduce it to rubble with wave upon wave of costly and dangerous bombing attacks. Fairchild's point was, Why would you do that if you could use your intelligence, and the magic of the Norden bombsight, to disable a city with a single strike? As Pape told me:

> They're certainly thinking that the bomber alone or airpower alone is going to win the war. And what they're thinking is that it's going to win the war and prevent a mass carnage like what occurred in World War I, where the armies clashed together year after year after year, and millions and millions of people died in the meat grinder of the trenches.

You can see why Donald Wilson, only half jokingly, said that if the Army had known what was going on at Maxwell, they would have put all the members of the Bomber Mafia in jail. Because these men were *part* of the Army, but they were saying that the rest of the Army was irrelevant and obsolete. You could have hundreds of thousands of troops massed along the Canadian border, complete with artillery and tanks and every other weapon imaginable, but the bombers would just fly right over them, leapfrog all the conventional defenses, and cripple the enemy with a few carefully chosen air strikes hundreds of miles beyond the front lines.

Tami Biddle, a professor of national security at the US Army War College, explains the Bomber Mafia's psychology this way:

> I think there's a fascination with American technology. I think there's a strong moral component to all this, a desire to find a way to fight a war that is clean and that is not going to tarnish the American reputation as a moral nation, a nation of ideas and ideology and commitment to individual rights and respect for human beings.

The Bomber Mafia—despite its ominous name—was never very large. It was a dozen men at most, all living—more or less—within walking distance of one another on those quiet, shaded streets at Maxwell Field. Nor was the Tactical School itself some massive facility. It was never West Point, churning out generation after generation of Army officers. During its twenty years of operation, it produced just over a thousand graduates. Had the Second World War never happened, it is entirely possible that the theories and dreams of this little group would have faded into history.

But then Hitler attacked Poland, and Great Britain and France declared war on Germany, and by the summer of 1941 it was obvious to everyone that the United States would soon be at war as well. And if the country was to go to war, it was obvious that it would need a strong air fleet. But what did a strong air fleet mean? How many planes did it need? To answer that question, the Army high command in Washington turned in desperation to the only group of experts who might have an answer: the instructors at the Tactical School, down at Maxwell Field in Alabama.

So the Bomber Mafia went to Washington and produced an astonishing document that would serve as a template for everything the United States did in the air war. The document is titled "Air War Plans Division One" (AWPD-1). It lays out, in exacting detail, how many planes the United States would need—fighters, bombers, transport planes. Also how many pilots. How many tons of explosives. And the targets in Germany for all those bombs, chosen according to the choke-point theory: fifty electrical power plants, forty-seven transportation networks, twenty-seven synthetic oil refineries, eighteen aircraft assembly plants, six aluminum plants, and six "sources of magnesium." And this astonishing set of projections was produced just in nine days, start to finish—the kind of superhuman feat that is only possible if you have spent the previous ten years in the seclusion of central Alabama, waiting for your chance.

The Bomber Mafia was ready for war.

CHAPTER THREE

"He was lacking in the bond of human sympathy."

A British motorcycle messenger drove up to my residence at Castle Combe, outside London. And the message that he delivered to me was from General [Hap] Arnold, which, when decoded, said, "Meet me tomorrow morning at Casablanca."

—Commanding General Ira Eaker

1.

Casablanca, in what was then French Morocco, was the site of a secret conference in January of 1943 between Winston Churchill and Franklin Roosevelt.

The war was just starting to turn in the Allies' favor, and the two leaders were meeting to plan what they hoped would be the final, victorious chapter. Both men brought their senior military brass. For Roosevelt that included General Hap Arnold, who commanded all American airpower. And now, midway through the conference, Arnold was sounding the alarm by sending an urgent wire to his most important deputy.

Ira Eaker was a distinguished graduate of the Air Corps Tactical School at Maxwell Field. Eaker was a charter member of the Bomber Mafia, a true believer in daylight high-altitude precision bombing. And he was the head of the Eighth Air Force—the fleet of bombers stationed in England that was charged with hitting all the targets outlined in the crucial war-planning document AWPD-1.

Come to Casablanca, the message to Eaker said. *Now*.

As Eaker recalled it:

They had kept the Casablanca Conference under such secrecy and wraps that I didn't even know what that meant. But I knew that I'd better comply. So I called General [Frederick Louis] Anderson, who was the bomber commander, and I said, "Have one of your crews pick me up in a B-17 at Bovington at midnight tonight to fly me to Casablanca, to arrive there shortly after day-light tomorrow morning."

Eaker arrived and went straight to General Arnold's villa.

And General Arnold said, "I have bad news for you, son. Our president has just agreed, upon the urging of the prime minister, that we discontinue our daylight bombing and you join the RAF in night bombing."

The RAF was the Royal Air Force. The ideas that had so enthralled Eaker and his classmates at Maxwell Field did not have quite the same effect on the other side of the Atlantic. The British were skeptical about precision bombing. They had never fallen in love with the Norden bombsight. They never got tantalized by the possibility of dropping a bomb into a pickle barrel from thirty thousand feet. The Bomber Mafia said that you break the will of your enemy by crippling it economically—by carefully and skillfully taking out the aqueducts and the propeller-spring factories—so that the enemy is incapable of going on. They believed that modern bombing technology allowed you to narrow the scope of war. The British disagreed. They thought the advantage of having fleets of bombers was that you could *broaden* the scope of war. They called it "area bombing," which was a euphemism for a bombing strategy in which you didn't really aim at anything in particular. You just hit everything you could before flying home.

Area bombing is not done in daylight, because if you aren't bombing at anything specific, why do you need to see anything? And it was explicitly aimed at civilians. It said: You *should* hit residential neighborhoods, and keep coming night after night, in wave after wave, until your enemy's cities are reduced to rubble. Then the will of the enemy is going to sink so low that it will just give up. When the British wanted a better euphemism for what they were doing, they called it "morale bombing"—bombing with the intent to destroy the homes and cities of your enemy and reduce your enemy's population to a state of despair.

The British thought the American Bomber Mafia was crazy. Why were they taking all the risks of flying during the day against targets too hard to hit? The British were trying to win a war, and it seemed to them that the Americans were holding an undergraduate philosophy seminar.

So at Casablanca, Churchill said to FDR, *Enough. You're doing it our way now.* And in a panic, General Arnold summoned his commander in Europe, Ira Eaker, to tell him the bad news: area bombing had won the day.

But Ira Eaker was a member of the Bomber Mafia. He wasn't about to give up so easily.

In Eaker's words:

I said, "General, that makes no sense at all. Our planes are not equipped for night bombing; our crews are not trained in night bombing. We'll lose more crews coming back into this fog-shrouded island in the darkness than we will attacking German targets in the daytime." I said, "If they're going to make this kind of a mistake, count me out. I won't play." Well, he said, "I suspected that would be your reaction...I know the reasons you've outlined as well as you do. But...since you feel so strongly about it, I'll see if I can make a date for you to talk to the prime minister tomorrow morning."

Eaker went back to his quarters and stayed up half the night drafting a response for Churchill. Everyone knew that Churchill wouldn't read a document longer than a page. So the briefing had to be *really* brief. And convincing.

So when I reported in, the old PM came down the stairway—the high glass windows and the sun was shining through the orange groves—and he came down resplendent in his air commodore's uniform. He had a penchant, which I knew of— when he was seeing a naval person, he wore his naval uniform; air, air [uniform], and so forth. Well, he said, "General, your General Arnold tells me you're very unhappy about my request to your president that you discontinue your daylight bombing effort and join Marshal [Arthur] Harris and the RAF in the night effort." I said, "Yes, sir, I am. And I've set down here on a single page the reasons why I'm unhappy. And I have served long enough in England now to know that you will listen to both sides of any controversy before you make a decision." So he sat down on the couch and took up this piece of paper, called me to sit beside him, and he started reading. And he read like some aged person, with his lips, half audibly.

So what did Eaker write? The most basic argument he could come up with. "I'd said that if the British bombed by night and the Americans by day, bombing them thus around the clock will give the devils no rest." When he got to that point of the memo, Churchill repeated the line to himself. As if he were trying to understand the logic. Then he turned to Eaker.

He said, "You have not convinced me now that you are right, but you have convinced me you should have a further opportunity to prove your case. So when I see your president at lunch today, I shall say to him that I withdraw my objection and my request that you join the RAF in night bombing, and I shall suggest that you be allowed to continue for a time."

The Americans got a reprieve. By the skin of their teeth.

2.

Put yourself in the shoes of the Bomber Mafia at this moment: Ira Eaker, Haywood Hansell, Harold George, Donald Wilson, all the others from the Air Corps Tactical School. They have been working side by side with their closest ally to defeat the Nazis. And yet their ally seems incapable of comprehending the conceptual advance they have made in waging war.

When he first got to England, Eaker lived at the home of his counterpart in the Royal Air Force, Arthur Harris, otherwise known as Bomber Harris. They would drive together every morning to bomber command headquarters, at High Wycombe.

As historian Tami Biddle explains:

It's very odd. Ira Eaker and Arthur Harris have doctrines of bombing that are 180 degrees out from one another, completely different. Yet they become fast friends. They really genuinely like each other. In fact, at one point, Harris tells Eaker, if anything happens to [my wife] Jill and me…we'd like you to have [our daughter] Jackie. We'd like you to be her godfather. It's quite the interesting relationship, but they are operating in completely different ways.

Marshal Harris's steadfast belief in the power of "morale bombing" must have offended Eaker. Or at the very least baffled him. Because what had the British just been through? The Blitz. The Blitz was a textbook example of area bombing. On September 4, 1940, Hitler had declared: "The hour will come when one of us will break, and it will not be the National Socialist Germany!" And in the fall of 1940, he sent German bombers thundering across the skies above London, dropping fifty thousand tons of high-explosive bombs and more than a million incendiary devices.

Hitler believed that if the Nazis bombed the working-class neighborhoods of East London, they would break the will of the British population. And because the British believed the same theory, they were terrified that the Blitz would cost them the war. The British government projected that between three and four million Londoners would flee the city. The authorities even took over a ring of psychiatric hospitals outside London to handle what they expected to be a flood of panic and psychological casualties.

But what actually happened? Not that much! The panic never came.

As a British government film from 1940 described it, "London raises her head, shakes the debris of the night from her hair, and takes stock of the damage done. London has been hurt during the night. The sign of a great fighter in the ring is, Can he get up from the floor after being knocked down? London does this every morning."

The psychiatric hospitals were switched over to military use because no one showed up. Some women and children were evacuated to the countryside as the bombing started, but by and large people stayed in the city. And as the Blitz continued, as the German assaults grew heavier, the British authorities began to observe—to their astonishment—not just courage in the face of the bombing but also something closer to indifference.

The Imperial War Museums later interviewed many survivors of the Blitz, including a woman named Elsie Elizabeth Foreman. As she described it,

We used to go in the shelter all the time, and then as they petered off a little bit, we got a bit blasé, I suppose you might say. And we stayed in bed some of the time, but we still used to go dancing. [If] there was an air raid on, if anybody wanted to leave, they could, and all that. And the same at the pictures, if we went to the pictures…we used to just sit there. We never used to move and go out or anything until the actual time when we were bombed out twice, I think. We weren't actually bombed out the first time, just the glass…

One of my sisters—she came home and she was sweeping the glass from the front, because all the windows came in. But she swept it into the curb. And my eldest sister came out and—this was during an air raid that the all-clear hadn't gone. And they had this terrific row because my sister had put my oldest sister's best high-heel shoes on, which were very hard to get in those days, same as silk stockings were…Bombs were dropping all over the place, and there were these two having a row over a pair of shoes and sweeping the glass at the same time.

It turns out that people were a lot tougher and more resilient than anyone expected. And it also turns out that maybe if you bomb another country day in and day out, it doesn't make the people you're bombing give up and lose faith. Maybe it just makes them hate you, their enemy, even more. The area-bombing advocates had this cleverly deceptive word they used to describe the effect of their bombing: *dehousing*. As if you could destroy a house without disturbing its occupants. But if my house is gone, doesn't that make me more dependent on my government, not more inclined to turn on my government?

Historian Tami Biddle takes the long view on area bombing: "I think we've seen this over and over again in the history of bombing. We've seen [that] the state, the target state—if we're talking about coercive bombing, long-range coercive bombing—finds ways of absorbing the punishment if it's really determined to do so."

When Blitz survivor Sylvia Joan Clark was asked whether she ever thought the Germans might win the war, she replied,

> No. I never thought that. I am very proud to be English, and I thought they'll never beat us. Never. I had that in my heart that if I worked, and I helped everybody, we'd get there in the end...I used to say this to people. It's no use being down. I had a home. I've had a mother. I've had a father and I've lost [them], but I've made up my mind nobody's going to get me down. I'm going to survive, and I'm to work hard and be proud that England will be England again.

Once they tallied up the damage, the British determined that more than forty-three thousand people had been killed and tens of thousands injured. More than a million buildings were damaged or destroyed. And it didn't work! Not on London or Londoners. It did not crack their morale. And despite that lesson, just two years later, the Royal Air Force was proposing to do the exact same thing to the Germans.

Ira Eaker said that he and RAF Marshal Harris, when they were living together, had discussions—though I'm guessing *arguments* would be a better term. They'd talk long into the night, and once, Eaker turned to Harris and made this exact point: "I asked Harris if the bombing of London had affected the morale of the British. He said it made them work harder. But in the case of the Germans, however, he thought the reaction was different because they were a different breed from the British."

To Eaker and the rest of the Bomber Mafia, the British attitude made no sense. And it was only later that they came to understand why. The British had their own version of a Bomber Mafia—with an equally dogmatic set of views about how airpower ought to be used. Actually, the word *mafia* is not quite right—more like a single bombing mafioso. A godfather. And his name was Frederick Lindemann.

3.

In the decades after the Second World War, scholars on all sides tried to make sense of what the war had meant, and among them was a prominent British scientist named C. P. Snow. Snow had served in the British government during the war. He was a Cambridge don, a successful novelist, and friends with everyone who was anyone in the British intellectual elite. In 1960 he came to Harvard University to give a lecture, a big chunk of which was devoted to the story of Frederick Lindemann.* Snow believed that Lindemann had played a hugely underappreciated role in the way the British chose to use their airpower. If you wanted to understand the befuddling attitude the British had about bombing, Snow said, you had to understand Lindemann.

As Snow put it in his Harvard lecture:

Lindemann was by any odds a very remarkable and a very strange man. He was a real heavyweight of personality…

Lindemann was quite un-English. I always thought if you met him in middle age, you'd have thought he was the kind of central European businessman that one used to meet in the more expensive hotels in Italy…

I mean, he might have come from Düsseldorf. He was heavy-featured, pallid, always very

---

* I explore more about Lindemann in "The Prime Minister and the Prof," an episode from the second season of my podcast, *Revisionist History*.

correctly dressed. He spoke German at least as well as he did English, and indeed under his English there was a tone of German—if you could hear him at all, because he always mumbled in an extraordinarily constricted fashion.

Frederick Lindemann—later known as Lord Cherwell—was born in Germany in 1886. His father was a wealthy German engineer. His mother was an American heiress. Lindemann was a physicist and got his PhD in Berlin just before the First World War—at a time when Germany was the center of the world in physics. Colleagues compared his mind to Isaac Newton's. He had an extraordinary memory for numbers: as a child, Frederick would read newspapers and recite back reams and reams of statistics from memory. He could demolish anyone in an argument. He also spent a considerable amount of time with Albert Einstein. Once, at dinner, Einstein mentioned some mathematical proposition for which he'd never been able to come up with a proof. The next day Lindemann casually mentioned that he had the answer; he'd figured it out in the bathtub.

Everyone talked about Lindemann. And for a writer like Snow, the gossip was irresistible.

His passions were much bigger than life…[They] reminded me…of the sort of inflated monomania of the passions in Balzac's novels. He'd have made a wonderful Balzacian character. And, I said, he's a figure who made a novelist's fingers itch.

He enjoyed none of the sensual pleasures. He was the most cranky of all vegetarians. He wasn't only a vegetarian, but he would only eat very minute fractions of what you might regard as a vegetarian diet. He lived mainly on Port Salut cheese, the whites of eggs—the yolks being apparently too animal—olive oil, and rice.

Lindemann was eccentric and brilliant. But his greatest claim to fame was that he was Winston Churchill's best friend. The two men had met each other in 1921 at a dinner arranged by the Duke and Duchess of Westminster. Churchill was an aristocrat, and Lindemann was really rich. So the two moved in the same circles. They hit it off. As for Churchill, if you read some of the letters he wrote Lindemann, they are almost worshipful.

The psychologist Daniel Wegner has this beautiful concept called transactive memory, which is the observation that we don't just store information in our minds or in specific places. We also store memories and understanding in the minds of the people we love. You don't need to remember your child's emotional relationship to her teacher because you know your wife will; you don't have to remember how to work the remote because you know your daughter will. That's transactive memory. Little bits of ourselves reside in other people's minds. Wegner has a heartbreaking riff about what one member of a couple will often say when the other one dies—that some part of him or her died along with the partner. That, Wegner says, is literally true. When your partner dies, everything that you have stored in that person's brain is gone.

Churchill's personality is important here. He was a man of the big picture. A visionary. He had a deep, intuitive understanding of human psychology and history. But he struggled with depression. He had mood swings. He was impulsive, a gambler. He had no head for figures. Throughout his life he was always losing huge amounts of money on foolish investments. In 1935, Churchill spent the modern equivalent of more than $60,000 on alcohol—in one year. Within a month of becoming prime minister, he was broke.

Here we have a man with very little common sense, no ability to handle numbers, no way to bring order to his life. And so whom does he become best friends with? Someone disciplined, almost fanatically consistent. Someone who ate the same three things at every meal, every day. Someone so naturally at home in the world of numbers that, as a child, he would read newspapers and recite back reams and reams of statistics from memory.

Churchill stored all the thinking that had to do with the quantitative world inside Lindemann's brain. And when Churchill became prime minister, in 1940, just after the war broke out, he took Lindemann with him. Lindemann served in Churchill's cabinet as a kind of gatekeeper to Churchill's mind. He went with Churchill to conferences. He dined with him. Lindemann never drank unless he was eating with Churchill, who was a big drinker. Then he drank. He went to Churchill's country house on the weekends. People spotted them at 3:00 a.m., sitting by the fire, reading the newspaper together.

As Snow put it, "It was an absolutely true and very deep friendship, and both men paid some price for it. And when Lindemann was very much disliked by other of Churchill's intimate associates, Winston never budged. They tried to get rid of Lindemann, but Churchill wouldn't have it."

One of the subjects on which Lindemann was most persuasive, when it came to Churchill, was bombing. Lindemann was a great believer in the idea that the surest way to break the will of the enemy was by bombing its cities indiscriminately. Now, did Lindemann have any evidence to support his idea? No. In fact, that was the whole point of C. P. Snow's lecture—to show that this man of science, this brilliant intellectual, manufactured and distorted the facts to support his case:

No one had ever thought how these bomber forces were really to be used. It was just an act of faith; this was a way to fight a war. And I think it's fair to say that Lindemann was, with his usual extreme intensity, as committed to this faith as any man in England. Early in 1942 he was determined to put it into action.

In America, at the Air Corps Tactical School, the Bomber Mafia dreamed of a world where bombs were used with dazzling precision. Lindemann went out of his way to promote the opposite approach—and the only explanation Snow could come up with is personal. Lindemann was just a sadist. He found it satisfying to reduce the cities of the enemy to rubble: "About him there hung a kind of atmosphere of indefinable malaise. You felt that he didn't understand his own life well, and he wasn't very good at coping with the major things. He was venomous; he was harsh-tongued; he had a malicious, sadistic sense of humor, but nevertheless you felt somehow he was lost."

One of Lindemann's biographers once wrote of him: "He would not shrink from using an argument which he knew to be wrong if by so doing he could tie up one of his professional opponents."

And here's what a friend said of him: "He was indeed lacking in the bond of human sympathy for every chance person who was not brought into a personal relationship with him." One time Lindemann was asked for his definition of morality, and he answered: "I define a moral action as one that brings advantage to my friends."

Well, there you are. *I define a moral bombing action as one that brings advantage to my friend Winston Churchill.* So Lindemann writes Churchill one of his famous memos. As Snow described the document:

> It was a paper suggesting that every resource in England should be used to make bombers, to train bombing crews, to use all these bombers and bombing crews on the bombing of German working-class houses. It described in quantitative terms the results of a bombing offensive...The calculation was that if you gave total effort, you could destroy half the working-class houses in all the big towns in Germany. That is 50 percent of the...towns with populations over fifty thousand within the period of eighteen months. Fifty percent of the houses, according to Lindemann, would no longer exist.

So Lindemann convinced Churchill. And Churchill appointed Arthur Harris—the man whose home Ira Eaker stayed at when he first came to England—to run the British bombing command. And Arthur Harris was a psychopath. His own men called him Butcher Harris.

In one of his first major statements upon taking the post, Harris quoted Hosea, one of the bleakest of the

Old Testament prophets: "The Nazis entered this war under the rather childish delusion that they were going to bomb everyone else and nobody was going to bomb them…They sowed the wind, and now they are going to reap the whirlwind."

Shortly after taking over British bombing operations, Harris launched a massive attack on the city of Cologne. A night bombing, because of course they didn't particularly need to see their targets, did they? Harris sent one thousand bombers into Germany, and they dropped their bombs everywhere. In the end, the RAF campaign leveled 90 percent of central Cologne, six hundred acres in all. More than three thousand homes were destroyed.

Once, during the war—the story goes—Harris was stopped for speeding. The policeman said, "Sir, you are traveling much too fast; you might kill someone." Harris replied, "Now that you mention it, it's my business to kill people: Germans."

Years later, in 1977, Harris was interviewed by the British Forces Broadcasting Service. He'd had more than thirty years to think about his actions.* But when he spoke about one of his most infamous missions,

---

* In 1969, Kurt Vonnegut published his novel *Slaughterhouse-Five*. Although it is framed as science fiction, the novel is in large part based on Vonnegut's experience as an American POW in Dresden during the RAF bombing campaign. The novel stayed on the *New York Times* bestseller list for sixteen weeks.

when his bombers reduced the city of Dresden to rubble, there was no remorse:

> Well, of course people are apt to say, "Oh, poor Dresden, that lovely city. Solely engaged in producing beautiful little china shepherdesses with frilly skirts." But as a matter of fact, it was the last viable…governing center of Germany. And also, it was virtually the last way through from north to south for German reserves, moving in front of the Russian, and our own, army advances.

Ostensibly to prevent the movement of troops through Dresden, Harris had his bombers take out 1,600 acres in the city's core and kill twenty-five thousand civilians over the course of three days. When asked why he targeted civilians instead of military installations, Harris challenged the question:

> We weren't aiming particularly at the civilian population. We were aiming at the production of everything that made it possible for the German armies to continue the war. That was the whole idea of the bombing offensive. Including, as I said, the destruction of the facilities for building submarines and the armament industries throughout Germany and the people who worked in them. They were all active soldiers, to my mind. People

who worked in the production of munitions must expect to be treated as active soldiers. Otherwise, where do you draw the line?

They were all active soldiers, to my mind. Children. Mothers. The elderly. Nurses in hospitals. Pastors in churches. When you make the leap to say that we will no longer try to aim at something specific, then you cross a line. Then you have to convince yourself that there is no difference between a soldier on the one hand and children and mothers and nurses in a hospital on the other.

The whole argument of the Bomber Mafia, their whole reason for being, was that they didn't want to cross that line. They weren't just advancing a technological argument. They were also advancing a moral argument about how to wage war. The most important fact about Carl Norden, the godfather of precision bombing, is not that he was a brilliant engineer or a hopeless eccentric. It's that he was a devoted Christian.

As historian Stephen McFarland puts it,

You might wonder, if he thought he was being in service to humanity, why he would develop sights to help people drop bombs. And the reason was because he was a true believer that by making bombing accuracy better, he could save lives.

He truly believed what the Army and Navy were telling him. And that is that we're going to destroy machines of war, not the people of war. We're not going to do like [we did in] World War I, where we slaughtered millions of soldiers. We're not going to try to slaughter millions of civilians. We're only going to try to blow up factories and blow up machines of war. And he bought into that. That was part of his basic philosophy of life, his Christianity.

So for Commanding General Ira Eaker, that midnight trip to Casablanca to save precision bombing was the most morally consequential act of his life. And when he came back to his air base in England, he said, *We need a new plan for the war in Europe, one that will show the British that there is a better way to wage an air war.* And whom did he pick to think up that plan? Haywood Hansell, now General Hansell, one of the brightest of the young lights in the US Army Air Forces. The same Hansell who would one day abruptly lose his job to Curtis LeMay on the island of Guam.

CHAPTER FOUR

"The truest of the true believers."

1.

Haywood Hansell came from an aristocratic southern military family. His great-great-great-grandfather
John W. Hansell served in the American Revolution.
His great-great-grandfather William Young Hansell
was an Army officer in the War of 1812. His great-
grandfather was a general in the Confederate Army,
his grandfather a Confederate officer. And his father
was an Army surgeon who came to dinner in a
white linen suit and a panama hat. Haywood liked
to carry a swagger stick, as the British Army officers did. Everyone called him Possum, his childhood
nickname.

Hansell was slender and short—a skilled dancer, a poet, and an aficionado of Gilbert and Sullivan operettas. His favorite book was *Don Quixote*. He put flying first, polo second, and family a distant third. Once, early in his marriage, the story goes, he heard a baby cry and turned to his wife. "What in heaven's name is that?" "That's your son," she said. On his final combat mission as a pilot, a bombing run over Belgium, Hansell entertained his exhausted crew with a rendition of the popular music-hall song "The Man on the Flying Trapeze." As C. P. Snow would have put it, Hansell is the kind of character who makes a novelist's fingers itch.

In wartime, combat units are obliged to inform the press of their accomplishments, so that the folks back home can learn of the progress of the war. But military press releases tend to be loaded with so many euphemisms, elaborations, and aggressive improvements on the truth that if placed in any body of water, they would sink immediately to the bottom. By contrast, consider a press release from December of 1944, personally dictated by Hansell from his headquarters on Guam. He wrote: "We have not put all our bombs exactly where we wanted to put them, and therefore we are not by any means satisfied with what we have done so far. We are still in our early experimental stages. We have much to learn and many operational and technical problems to solve."

We have much to learn. That's Hansell: un-flinchingly honest, a little naive, but fundamentally a romantic, with all that implies. Once, while posted at Langley Field, in Virginia, he passed by a young woman in the lobby of a hotel—Miss Dorothy Rogers of Waco, Texas. Hansell immediately took his own date home, returned to the hotel, and invited himself to join the young woman and her aunt for dinner. Dorothy Rogers found him tiresome. He found her delightful. She returned to Texas. He wrote her every day for the better part of a year. She answered two, maybe three of his letters. They were married in 1932.

It stands to reason that Hansell's favorite book was *Don Quixote.* Don Quixote is the gallant knight distin-guished by his ceaseless, courageous crusade to revive chivalry. Don Quixote tilted at windmills, suffered endless deprivations, battled imaginary enemies. Don Quixote would have written a woman he barely knew hundreds of times, even as she all but ignored him. But Quixote is a strange choice for a military man, isn't he? The don holds to an ideal, but that ideal is never realized. It's based on an illusion. He thinks he is making the world a better place, but he actually isn't. Consider this passage from *Don Quixote,* which Haywood Hansell, in his long years of retirement after the humiliation of Guam, may well have read—and winced in self-recognition:

In short, [Don Quixote] became so absorbed in his books that he spent his nights from sunset to sunrise, and his days from dawn to dark, poring over them; and what with little sleep and much reading his brains got so dry that he lost his wits. His fancy grew full of what he used to read about in his books, enchantments, quarrels, battles, challenges, wounds, wooings, loves, agonies, and all sorts of impossible nonsense; and it so possessed his mind that the whole fabric of invention and fancy he read of was true, that to him no history in the world had more reality in it.

There's more than a little bit of Haywood Hansell in that.

In 1931, as a young Army lieutenant, Hansell was assigned to Maxwell Field. He was appointed an instructor at the Air Corps Tactical School in 1935 and distinguished himself quickly as one of the sharpest minds in the entire school. When Ira Eaker was looking for someone to defend the doctrine of high-altitude daylight precision bombing against the skepticism of the British, there was no question whom he would pick. That was a job for Haywood Hansell, the truest of the true believers.

2.

In a talk he gave in 1967, Hansell described the first problem he faced: "The selection of the targets themselves was a pretty complicated affair, an effort to gauge the effect of the destruction of a particular industry upon the war-making capacity of Germany."

Hansell needed to find a target the American bombers in England could easily reach and easily destroy. Something so critical to the Nazi war effort that the Germans would *suffer* if they lost it. And it had to be something specific. It wouldn't make any sense to target, say, railway bridges over the Rhine, the central waterway of Germany. There are dozens and dozens of railway bridges over the Rhine, spread out over hundreds of miles. Trying to hit them all would be a logistical nightmare.

Then Hansell heard about what happened after the Germans bombed a Rolls-Royce aircraft engine plant in the English city of Coventry. The attack was only partially successful, but it blew out the building's skylights, opening the factory floor to the elements. As he described it, "There was a rain, and thousands of trays of ball bearings were rusted and they couldn't be used. Engine production stopped at a time when it was desperately needed. It became quite apparent that the rotating machinery was extremely sensitive to the ball-bearing industry."

Hansell wondered whether ball bearings might be the Achilles' heel of Germany.

Why ball bearings, specifically? Because they are at the heart of any mechanical device. Tiny metal balls covered in grease and encased in a steel ring. Inside the axle of a bicycle, for example, there are perhaps a dozen ball bearings, acting as mini steel rollers that allow the bicycle wheel to turn freely. A good road bicycle can cost thousands of dollars and includes some extraordinarily sophisticated space-age materials. But without two or three dollars' worth of quarter-inch-diameter ball bearings, the bike won't work. It literally won't move. Same is true for the engine in your car. Or virtually any mechanical object that involves a rotating part.

Ball bearings were a huge issue for Carl Norden when he was building his first prototypes. The bombsight was a mechanical computer made up of dozens of moving parts, each of which had to rotate precisely to the right position in order for the calculations of the bombsight to be accurate. So if he had ball bearings that were of unequal sizes, or weren't completely smooth, the whole bombsight would be thrown off.

Historian Stephen McFarland explained how Norden addressed the issue: "[He] paid dozens of people to spend a day—or two or three—polishing a ball bearing. They would measure it every twenty seconds to make sure that it was absolutely round."

The problem, McFarland says, was that when the war started, Norden suddenly had to make thousands of bombsights. Which meant he couldn't have his ball bearings hand-polished anymore.

So Barth, his partner, who was the production guy, came up with a very interesting idea. He would come to a company and say, "I want you to produce hundreds of thousands of ball bearings." He then paid people to measure each ball bearing. And when they found a perfect ball bearing or one that met tolerances, that would be the one that would go into the bombsight. And they might have to look through fifty, sixty, a hundred other ball bearings, and they would throw them out because it was much cheaper that way.

Ball bearings were crucial to everything in modern warfare. And where was the German ball-bearing industry located? It turns out nearly all of it was concentrated in a medieval Bavarian town called Schweinfurt. Five separate factories, operating around the clock, employing thousands of people, supplied the German war machine with millions of ball bearings a month.

Schweinfurt was a Bomber Mafia fantasy. In the words of Tami Biddle,

If you took out that target, it could have the potential to take down the entire German war economy. This is what the Americans were looking for, and they thought ball bearings might be that target.

It's sort of like taking the key card out of a house of cards and having the whole thing collapse, or pulling on the thread of a spiderweb and having the whole thing unravel. That's what the Americans thought they were going to do. Again, it's very ambitious. It's resting on assumptions that were unproven, but very hopeful.

The Army Air Force strategists drew up one of the most ingenious plans of the war: a raid in two parts. The main event would involve 230 B-17 bombers sent against the Schweinfurt ball-bearing factories.

But to make the main event possible, there was to be a diversion. Just before the B-17s left for Schweinfurt, another fleet of B-17s would take off for Regensburg, a small city southeast of Schweinfurt. The Germans made their Messerschmitt fighter plane there. The idea was that the Regensburg attack would draw off the German defenders—occupy them, distract them—leaving a clear path for the bomber group headed for Schweinfurt. The bombers heading for Regensburg would be bait.

And whom did they choose to command this crucial, treacherous second arm of the Schweinfurt raid? The best combat commander they could find: a young Army Air Forces colonel named Curtis Emerson LeMay.

3.

Curtis LeMay came from a poor neighborhood in Columbus, Ohio—the eldest of a large family that struggled financially. He put himself through engineering school at Ohio State, working night shifts at a foundry. He joined the Army out of college—and his ascent through the Air Corps was breathtaking. A captain by thirty-three, then a major, a colonel, a brigadier general, and by the age of thirty-seven a major general.

LeMay was a bulldog. He had an oversize square head, with hair parted triumphantly just a shade off the middle. He was a brilliant poker player. A crack shot. He had a mind that moved only forward, never sideways. He was rational and imperturbable and incapable of self-doubt.

Consider this transcript of an interview from 1943. LeMay is in England, heading up the 305th Bombardment Group. He's just landed after taking his men on a bombing run.

Question: Colonel LeMay, how'd the trip go today?
LeMay: Well, it went pretty well, except it was
rather dull compared to some that we've had.

There weren't any fighters out, and flak was just moderate and very inaccurate.

A film crew had come to interview his airmen after the mission. The rest of the men are laughing, excited. A film crew! A chance to shine. LeMay—short, barrel-chested, pugnacious—looks, expressionless, at the camera. That raid deep in enemy territory? *It was rather dull.*

Question: This formation that you put out for us last night—did you conform to that on your trip, then?

LeMay: Yes, we flew the same formation we'd imagined last night.

Question: How about your bombardier—was he operating all right?

LeMay: He worked 100 percent as usual. [*laughter*]

Question: Major Preston here—did he perform his duties properly?

LeMay: Yes, he was right on the ball, same as he always is.

LeMay speaks with no inflection. No elaboration. It is safe to say that Colonel LeMay did not serenade his men with "The Man on the Flying Trapeze."

Question: How about the men—did they perform their duties?

LeMay: Crew's right up to par.

Question: You don't have any complaints, in other words.

LeMay: No complaints at all.

No complaints at all. Curtis LeMay was not the sort to complain—not to an outsider, anyway. Had the film crew interviewed Haywood Hansell, he would have waxed eloquent, dropped a few smart remarks at his own expense, then invited everyone back to his officer's quarters for a drink. Hansell was the anti-LeMay.

When Hansell was at Maxwell Field, before the war, he was part of a group of daredevil pilots led by the flying ace Claire Chennault. They performed impossibly dangerous stunts in planes that were not designed for that kind of adventure. As Hansell himself would admit, it's a miracle he survived. Hansell *would* join a daredevil group. It suited his romantic flair. LeMay? He was the opposite of romantic.

Russell Dougherty, one of LeMay's fellow Air Force generals, loved to tell a story about a time, much later, when LeMay was briefed about a new airplane called the FB-111:

The briefings lasted about two and a half days…And finally, they wrapped up the briefing, and LeMay hadn't said a word the whole time. He was just sitting there…After they got all through, General LeMay said, "Is that it?" "Yes, sir! That's it." And he got up, and he says, "It ain't big enough," and he walked out. That was his only comment.

A two-and-a-half-day briefing, dismissed with four words.

In the fall of 1942, LeMay came to Britain with the Eighth Air Force. He headed up a squadron of B-17 bombers based out of Chelveston. And he made his mark immediately.

Here's one example: If you fly a fleet of B-17 bombers deep into enemy territory in order to precision-bomb from twenty thousand feet, how do you protect yourself from enemy fighter planes? Bombers had guns and armor plating, but it quickly became obvious, once the shooting started, that that still wasn't enough. So LeMay devised something called the combat box formation—a way for a group of bombers to fly together so that they could most easily defend themselves against enemy attack. It was an idea quickly adopted by the whole Eighth Air Force. Then LeMay turned his attention to an even bigger problem: his pilots.

As LeMay put it in an oral history long after he'd retired: "One of the things that was very apparent was that the bombing was not very good."

Bombers have cameras that take pictures, called

strike photos, of the area where their bombs fall. And when LeMay looked at the strike photos after the crews had come back to base, he could see that the bombs were landing everywhere *but* the target. "Not only were the targets not being destroyed, but we didn't have any records of where most of the bombs actually fell. They were taking strike photos, of course, but you could not locate over half of the bombs that were hauled over to the Continent."

The problem was that the pilots were not flying straight at the targets. They believed that would make them sitting ducks for antiaircraft fire, because enemy artillerymen on the ground would simply estimate the planes' speed and altitude and aim accordingly. So the pilots were taking evasive action, not flying directly at the target until the last seconds of their bombing run. Which is why the bombs were falling wide. How could the bombardier working the bombsight do his job if the plane was lined up over the target only at the very last moment?

LeMay explained, "Something had to be done to give the bombardier a chance to hit the target. This meant a longer bomb run to give him ample time to get the bombsight level."

LeMay saw only one solution. The pilots had to *stop* taking evasive action. They had to fly straight in, over the target. This went directly against received wisdom. "All of the people that I talked to that had been in combat were of the opinion that if you did this, antiaircraft guns would shoot you down," he said.

But that was just opinion. LeMay was an empiricist. He went back and studied his old artillery manuals and did some calculations. How many rounds from an antiaircraft gun would it take to bring down a B-17 bomber? As he recalled, "It required I think 377 rounds to hit it. This didn't look too bad to me."

An antiaircraft gun would have to fire 377 rounds if it hoped to disable a B-17 bomber flying straight at the target. Three hundred seventy-seven rounds is a lot of ammunition, so flying straight is a risk, but it's not a crazy risk.

So LeMay said, *Let's try it. Let's fly in straight. A seven-minute-long, straight and steady approach.* And if that sounded suicidal—which it did to all his pilots— he added, *I'm going to be the first to try it.* In a 1942 bombing run over Saint-Nazaire, France, LeMay led the way. He took no evasive action. And what happened? His group put twice as many bombs on the target as any group had before. And they didn't lose a single bomber.

Robert McNamara, who later became secretary of defense during the Vietnam War, ran analysis for the Army Air Forces during World War II. In Errol Morris's brilliant documentary *The Fog of War,* McNamara described LeMay after he heard that so many pilots were turning tail:

He was the finest combat commander of any service I came across in war. But he was extraordinarily belligerent, many thought brutal. He issued an order. He said, "I will be in the lead plane on every mission. Any plane that takes off will go over the target, or the crew will be court-martialed." Now, that's the kind of commander he was.

The Bomber Mafia was made up of theorists, intellectuals who conceived of their grand plans in the years before the war from the safety of Montgomery, Alabama. But Curtis LeMay was the one who figured out how to realize those theories.

As LeMay said about the bombing mission that did away with evasive action: "I'll admit some uneasiness on my part and some of the other people in the outfit when we made that first straight-in bomb run, but it worked."

I'll admit some uneasiness, he says. That's it!

4.

One more LeMay story, because the fascination people have with LeMay—okay, the fascination I have with LeMay—is not that he was an extraordinary combat commander. There were plenty of those in the Second

World War. The fascination comes from the unfathomable depths of his character—the sense that he didn't have limits the way normal people did, which was, in one way, exhilarating, because it meant that LeMay could achieve things others could not even imagine. But at the same time, it gave people pause. Think about the word McNamara used to describe LeMay: *brutal.* And it's not like McNamara himself was warm and fuzzy. He would later direct the saturation bombing of North Vietnam. Yet LeMay gave *him* pause.

The story that started all the whispers about LeMay in military circles happened back in 1937, when the possibility of war in Europe was growing real. The Army Air Corps wanted a chance to practice their bombing technique. Real-world practice, only with dummy bombs: fifty-pounders filled with water. LeMay would talk about this exercise years later: "The Air Force has been battling to make a contribution to the defense of the country ever since I've been around. Nobody paid much attention to it…We wanted an exercise where we would drop bombs on a battleship. Find the battleship."

For the practice run to work, the Army Air Corps needed the Navy to play along. Hide a battleship out on the seas. Give out its coordinates at the last minute and dare the bombers to find it. This was before sophisticated radar and navigation aids. To find a battleship, you had to see it with your eyes, then hit its narrow decks with a bomb, from thousands of feet up—all while flying at hundreds of miles an hour.

The Navy was not enthusiastic.

"Finally they agreed that they would have an exercise. And it would be in August off the West Coast. Now, in August off the West Coast there's nothing but fog for a thousand miles out there. And they deliberately picked it at this time, I'm sure," LeMay said.

How could you spot a battleship in a thousand miles of fog? To make matters worse, the Navy bent the rules. The agreement was to have the war game run for twenty-four hours—from noon the first day until noon the next. But the Navy didn't give out the coordinates of its ship—the USS *Utah*—until late the first afternoon. And the coordinates they gave were wrong. They were off by sixty miles. One thousand miles of fog. Late directions. Fake directions. A needle in a haystack would have been easier to find.

At ten minutes before noon—at the very last moment—LeMay found the ship and dropped his bombs. Now, of course he found the ship. There was nothing LeMay could not do if he put his mind to it. That's not the point of the story. The point is what was going on just before he dropped his bombs.

The Navy was certain the ship couldn't be found, so it took no precautions. The sailors were just going about their business. They were supposed to take cover in a bombing exercise. They didn't.

What did LeMay do? He bombed the *Utah* anyway, raining fifty-pound water bombs down on the sailors.

As LeMay recalled, "Everybody [was] diving for the gangplanks, hatches. And we heard rumors that there were a few people hurt a little bit."

In his memoirs, LeMay says he heard that some sailors actually got killed in the bombing exercise, and then he writes, "I remember watching the first bomb, which smashed into the deck. It sent splintered pieces of wood flying in every direction. I hadn't realized that wood could frag like that."

He shrugs it off. His job was to find the ship, after all. And he did. And by the way, really good to know about the physics of a bomb hitting a wooden deck.

Conrad Crane, chief of historical services for the Army Heritage and Education Center, at Carlisle Barracks, and former director of the US Army Military History Institute, calls LeMay the greatest air commander in history:

He was a dynamic leader: he shared the difficulties of his airmen. He was the best navigator the Air Force had; he was a great pilot; he could do mechanic stuff. He knew the technical as well as the leadership aspects of what he was doing. He was the Air Force's ultimate problem solver.

But he was one of those guys that, if you gave him a problem to fix, you didn't ask a whole lot of questions how he was going to do it.

So imagine, then, the thinking of the Bomber Mafia in the summer of 1943. The men needed to validate the theories formulated back at the Air Corps Tactical School. They needed to deal a death blow to the Nazi war machine. They needed to prove that ball bearings were the crucial choke point of the German military infrastructure. The Schweinfurt raid was their best chance to demonstrate that their way of waging an air war was superior to that of the British. Whom would you choose to *plan* the mission? Haywood Hansell, of course, the high priest of Maxwell Field—one of your very best. But whom would you choose to *lead* the most difficult part of the mission—the dummy raid on Regensburg? There really wasn't any other option.

In a film entitled *The Air Force Story,* the narrator describes the scene: "Dawn, August seventeenth, 1943. England…The Eighth Bomber Command prepared 376 B-17s for the two most critical targets on their list: the ball-bearing plants at Schweinfurt and the Messerschmitt aircraft factory at Regensburg, both deep in Germany."

The airmen's story is also told in the first person:

By the time we turned in our personal stuff, it was well understood that the projected doubleheader would bring on a large-scale and costly air battle. In chapels all over England, most of the men turned to their ministers, rabbis, or priests...And this day our double mission involved the deepest penetration ever attempted into Germany. And the largest bomber force to be dispatched to date.

CHAPTER FIVE

"General Hansell was aghast."

1.

The orders given to Curtis LeMay on the eve of the Schweinfurt raid called for him to lead an elaborate decoy mission. He would take off first with the Fourth Bombardment Wing—a fleet of B-17 bombers. And they'd head for the Messerschmitt aircraft factories in Regensburg.

The idea was that LeMay's group would tie up the Germans defending the Messerschmitt factories. And then they would keep going, through the Alps to North Africa, in the hopes of luring the German fighter planes as far away as possible from the corner of Bavaria where the ball-bearing factories were.

As LeMay later recalled, "We'd go in and hit Regensburg and go on out the Brenner Pass, and we wouldn't have to fight coming out. [We] would bear the brunt going in of the German fighter force."

Then the real bombing force, the First Bombardment Wing, would arrive.

As LeMay put it: "They would get in practically free because the German fighter force would be working against the [Fourth Bombardment Wing]...and then be on the ground reloading. But they'd have to fight going in and coming out."

LeMay being LeMay, long before the day of the attack, he worried about the weather. He was taking off from the base in England, the land of mist and fog. So in the weeks leading up to the raid, he had his crews practice blind takeoffs, day after day.

Sure enough, on the morning of the mission, August 17, the fog was terrible. He remembered, "It's stinking over England. As a matter of fact, we went out that morning, and they had to take lanterns and flashlights and lead the airplanes out from the hard stands at the end of the runway."

LeMay led his men off into the gloom. Once they entered occupied France, the German fighters started to emerge from behind the clouds, and LeMay's Fourth Bombardment Wing learned what it meant to fly headfirst into the heart of the German air defense.

One of LeMay's pilots, Beirne Lay, wrote an article for the *Saturday Evening Post* a few months later, describing the Regensburg raid. And it's harrowing.

A shining silver rectangle of metal sailed past over our right wing. I recognized it as a main-exit door. Seconds later, a black lump came hurtling through the formation, barely missing several propellers. It was a man, clasping his knees to his head, revolving like a diver in a triple somersault, shooting by us so close that I saw a piece of paper blow out of his leather jacket...Now that we had been under constant attack for more than an hour, it appeared certain that our group was faced with extinction. The sky was still mottled with rising fighters. Target time was thirty-five minutes away. I doubt if a man in the group visualized the possibility of our getting much farther without 100 percent loss.

Lay describes another plane in his group: it was hit six times. One twenty-millimeter cannon shell penetrated the right side of the airplane and exploded beneath the pilot, cutting one of the gunners in the leg. A second shell hit the radio compartment, cutting the legs of the radio operator off at the knees. He bled to death. A third hit the bombardier in the head and shoulder. A fourth shell hit the cockpit, taking out the plane's hydraulic system. A fifth severed the rudder cables. A sixth hit the number 3 engine, setting it on fire. This was all in one plane. The pilot kept flying.

The attacks went on for hours before they reached Regensburg. The only solace they had was the thought that they were making life easier for the real attack—the one poised to cripple the Nazi war machine.

Except: the carefully constructed decoy mission turned out not to be much of a decoy at all. LeMay's pilots had been able to take off in the soupy fog of that August morning because he'd trained them for just that challenge. He had drilled them, takeoff after takeoff. *Use your instruments only. Act as if you can't see anything outside.* But no other group commander did what LeMay did. The flight crews were exhausted from their long runs into Germany, devastated by the loss of their comrades. They were sleepless, anxious, spent. Do you know how hard it is for a commander to turn to his crews and say, "This morning, at 6:00 a.m., we're going to practice blind takeoffs because of the *possibility* of fog on some future mission"?

Only LeMay could do that. He was relentless, a stickler. He didn't care if his men were grumbling as he pushed them on what must have seemed like a pointless exercise. Meanwhile, was Haywood Hansell paying attention to this detail? No. He was back in Washington, thinking loftier thoughts.

So that morning the bombers of the First Bombardment Wing were stranded on the tarmac until the weather cleared. They were supposed to take off ten minutes behind LeMay. They actually took off *hours* behind LeMay, which gave the German defenders time to regroup and launch the same ferocious assault on the Schweinfurt raid as they had a few hours earlier on the Regensburg raid.

In the end, there were *two* bloodbaths that day.

As LeMay recalled, "I had 125 airplanes, and I lost twenty-four, I think, which is not bad. But we only had a one-way trip. I think the First [Bombardment Wing], coming in an hour later—the German fighters were landed and back up again in force, and they had to fight coming in and going out, too. They lost about fifty or sixty airplanes."

Those are *staggering* losses. An air force that launches raids like that on a regular basis would quickly put itself out of business.

Even in its official histories, the Air Force could not hide the disaster. The narrator of *The Air Force Story* put it like this:

Göring's Luftwaffe unleashed every trick. The B-17s suffered the most savage blows since the war began…Battles lost us more men and aircraft in a single day than then Eighth Bomber Command had lost in our first six months of operations over Europe. We who carried the war five hundred miles to the enemy's industrial heart knew better than anyone how expensive it was.

As we began to run into flak, our gunners could feel the entire German Air Force warming up. Flying in enemy territory, we felt like goldfish in a bowl, waiting for the attack.

Each bomber was now committed. No more evasive action until "Bombs away." At this time, the formations were most vulnerable to attack. It didn't matter. We had a job to do on Schweinfurt. We had four hundred tons of high explosives to deliver.

But at least the mission took out the ball-bearing plants of Schweinfurt, crippling the German war effort—right? Well, not really.

In the film, the bombardiers peer into their sights. The bomb-bay doors open. The bombs fall in cascading waves. Then we see Germany, far below, erupting with explosion after explosion. The narrator continues: "After getting eighty hits on the two main ball-bearing plants, we could defend ourselves again. At least to the extent of [taking] evasive action against flak and fighter attack. But the main idea now was to get home fast."

Two hundred and thirty bombers, each carrying eight to nine bombs—so let's say two thousand bombs dropped in total. And they get eighty hits. That doesn't sound like precision bombing, does it?

2.

The fundamental problem at Schweinfurt was not the botched execution of the battle plan, however. That was just a symptom. The *real* problem had to do with the mechanical cornerstone of the Bomber Mafia ideology: the Norden bombsight.

As it turned out, the bombsight did not behave in the real world the way it had in Carl Norden's laboratory or in military training films. I asked historian Stephen McFarland whether the bombsight worked if conditions were ideal. His reply:

> Well, in theory, yes, if you're talking about strictly a mathematical issue. But remember that when gears and pulleys are moving, they cause friction, and I don't care how much you polish the ball bearings, I don't care how perfect the tolerances, you're still going to run into the issue of friction. And the slightest little bit of friction means that your analog equivalent to that mathematical formula has been messed up. It doesn't work that way anymore.

The Norden bombsight was a mechanical object. If you built it by hand, you could make sure that every component fitted perfectly and every tolerance was exact. But when the war hit, the military needed tens of thousands of machines.

As McFarland explains, "Once it's out of the factory, oils will start to thicken. At twenty-five thousand feet, the temperature might be sixty degrees below zero. And the oils that are lubricating the gears and pulleys are going to thicken and therefore cause a little bit of friction."

Now imagine that temperamental device in the hands of a bombardier—some kid, fresh out of training school—on an actual bombing run.

McFarland continues:

People are shooting at you, and enemy aircraft are coming at you at closing speeds of five hundred, six hundred miles an hour, and all this horrible yelling and screaming and bombs going, explosions going off and everything else—the bombardiers tended to pucker, if I can use that phrase. They would lean forward as they became more and more intent on trying to make sure the crosshairs stayed on the target. And when they did so, they actually changed the angle of vision through that telescope...It was impossible.

And I haven't mentioned the most important factor of all: the weather. The Norden depended on visual sighting of the target. You looked through the telescope, saw what you wanted to hit, then entered all the information: wind direction, airspeed, temperature, the curvature of the earth, and so on. But of course if there were clouds over the target, nothing worked. In the days before sophisticated radar, there was no way around this problem. You crossed your fingers and prayed for a sunny day. If you got clouds instead, sometimes you would scrub the mission.* But as often as not, you'd go anyway and take your chances. You had to. If you lingered too long on the tarmac, you would lose the element of surprise.

The Eighth Air Force took off in the fog for the ball-bearing factories of Schweinfurt. They dropped two thousand bombs. And of those, eighty found their mark. Eighty bombs are just not enough to destroy a sprawling industrial complex. When an employee of the Kugelfischer ball-bearing plant—one of the largest in the country—toured the factory after the attack, he found that the upper floor had completely collapsed. There was debris everywhere. But at least half the crucial machinery remained intact. Which meant that he could soon get it back up and running. Haywood Hansell thought he had found the classic choke point—the equivalent of that propeller-spring factory in Pittsburgh. But a plant that can be back up and running within a few weeks is not a choke point.

The best estimate was that the attack decreased German ball-bearing production by around a third.

---

* By the way, the same is still true today for many kinds of military drones. They need to see the target in order to aim at it.

Sixty planes and 552 airmen captured or dead for that? The Army's official postmortem of its bombing missions—the United States Strategic Bombing Survey—concluded afterward that "there is no evidence that the attacks on the ball bearing industry had any measurable effect on essential war production."

If this was the Bomber Mafia's attempt to prove the efficacy of its doctrine, it was a disaster, historian Tami Biddle says:

> The Americans were very outspoken about how much superior their method and their technique and their doctrine was, even when they had no grounds to be that bold and that confident, because they hadn't really proven anything.
>
> They hadn't done much. But they were, basically, cocky Americans who went into the theater thinking that the rules were going to be different for them and they were going to be able to do things that the British hadn't been able to achieve.

Yet what did the men of the Bomber Mafia do after the disaster of Schweinfurt? They tried again. In the fall of 1943, the Eighth Air Force hit Schweinfurt a second time.

A few years after the war, a movie came out called *Twelve O'Clock High.* It was based on a book written by Beirne Lay, the pilot under LeMay. *Twelve O'Clock High* starred Gregory Peck as the leader of an attack on a ball-bearing factory. It's worth watching because it perfectly captures the persistence of the Bomber Mafia's vision. The men had failed the first time, but it didn't matter. They would try again. Whatever evidence was slowly gathering about the limitations of the Norden bombsight didn't faze them. The dream was alive.

As the character General Pritchard, modeled after Ira Eaker, says in the film,

> There's only one hope of shortening this war. Daylight precision bombing. If we fold, daylight bombing is done with. And I don't know. Maybe it means the whole show. We could lose the war if we don't knock out German industry.
>
> You can smell what's coming, Frank. I'm promising you nothing except a job no man should have to do who's already had more than his share of combat. I've got to ask you to take nice kids and fly 'em until they can't take it anymore. And then put 'em back in and fly 'em some more.

What the movie doesn't do is follow the actual sequence of the first and second Schweinfurt raids—for obvious, Hollywood reasons. Because the second Schweinfurt raid was only marginally more successful than the first. It did more damage, but the German air-craft industry didn't grind to a halt that time, either. Not even close. And how many planes did the Eighth Air Force lose in that second raid? Sixty outright; seven-teen damaged so badly that they had to be mothballed; 650 airmen killed or captured. Nearly a *quarter* of the crews on that mission did not come home. Shortly thereafter, Ira Eaker—the leader of the Eighth—was reassigned. He was shunted over to the Mediterranean theater, which is the military equivalent of being sent to your room without dinner.

The year 1943 was a dark time for the Bomber Mafia. Every one of its ideas crumbled in the face of reality. The team was supposed to be able to put a bomb inside a pickle barrel from thirty thousand feet. That now seemed like a joke. And the bomber was supposed to fly so high and so fast that no one could touch it. *Are you kidding me?* US airmen of the Eighth Air Force were required to fly twenty-five missions to complete their tours of service. And if you were part of that second Schweinfurt mission, in which a quarter of the crews didn't come back—well, you do the math. Fly twenty-five missions like that, and what are your odds of making it through the war alive?

There are dozens of interviews from World War II airmen remembering those desperate months. One of those men, George Roberts, a B-17 radio operator with the Eighth Air Force, recalls:

We were assigned to a squadron, [the] 367th Bomb Squadron. And I noticed a big sign out there. It said this: HOME OF THE 367TH CLAY PIGEON AIR FORCE. Boy, and I thought, what a funny name, to call an outfit "clay pigeons." But…I was to find out later that "the clay pigeons" was a pretty good name for that squadron.

A clay pigeon is the name given to the targets used for shooting competitions: disks made out of clay, so that they shatter on impact, and colored fluorescent orange so they're hard to miss. That's not an encouraging name for a bombing squadron.

As the war over Europe dragged on, the pressure on the Bomber Mafia grew. The British became more contemptuous of the Eighth Bomber Command. Meanwhile, the brass back in Washington tried to push the air war in a new direction. They called for a different raid on Germany, an attack on the German city of Münster. Only Münster wasn't an industrial center. It didn't have an aircraft factory or a ball-bearing plant or an oil refinery. It was just a charming medieval town full of German civilians.

One pilot who flew the mission, Keith Harris, recalled,

We took off before the 390th on a mission to Münster, in Germany. It was on a Sunday, nice sunshiny day, beautiful day. Beautiful fall day. And the target was the built-up section of Münster. I thought it was rather inappropriate that these large set of steps in one big building in Münster was picked out as the aiming point.

He's talking about the Münster Cathedral. The Eighth Air Force was being directed to bomb a church on a Sunday at midday, as people were coming out of Mass.

At the preflight briefing, the airmen had been in shock. This wasn't what they had signed on to do. It wasn't what the Eighth Air Force stood for. One navigator—who had been raised in a strict Methodist household—went up to his commanding officer and said he couldn't do it. This was British-style area bombing, not American bombing. The navigator was told he faced court-martial if he didn't fly the mission. So he did. And you know who else was in that briefing room, trying to wrap his head around what was happening? Haywood Hansell. One of his airmen later wrote simply: "General Hansell was aghast."

3.

During the war, a young statistician named Leon Festinger worked on a project for the Army Air Forces.

His job was to devise better ways of selecting people for pilot training, which sounds like a dry academic exercise—until you remember how dire things were for the Air Forces in the long months of 1943. Festinger's job was essentially to figure out which young men should be sent to what—statistically speaking—was an almost certain death.

Leon Festinger went on to become one of the most famous social psychologists in the country. And I've always wondered whether his experience with the Air Forces was the motivation after the war for his most famous study, an analysis of a cult out of Chicago called the Seekers. Festinger approached the Seekers with a question that must have crossed his mind years before, during that dire period when everything the Bomber Mafia believed in was proved false: What happens to true believers when their convictions are confronted by reality?

As Festinger recalled, "The idea that you have to supply cognition that will fit with—that will justify— what you feel or what you do made this immediately the first thing we thought of: well, if this operates, it must be a very pervasive thing."

The leader of the Seekers was a woman named Dorothy Martin, who claimed to be in contact with a group of aliens she called the Guardians. The Guardians told her, she said, that the world was going to be destroyed by flood on December 21, 1954. But a few days before the apocalypse happened, she and her followers would be rescued by a flying saucer. It would land in her backyard. In preparation for this moment, the Seekers quit their jobs, left their families, and gave away their possessions. They gathered in Dorothy Martin's house, in the Chicago suburb of Oak Park. At first, Martin said, the flying saucer was supposed to arrive at four o'clock on December 17. The aliens didn't come. Then at midnight, Martin said she'd received a new message that the flying saucer was on its way. It never arrived. Then she said the aliens had given her a new date: midnight on December 21—just before the apocalypse. So the Seekers gathered again in Martin's living room and waited. And waited.

As Festinger recalled, "We were reasonably sure that their prediction was not going to be borne out. And so there we had a group of people who were committed to a certain prediction, and they were indeed committed. People had quit jobs, sold things. They were preparing for a cataclysm, for their personal salvation."

It's worth quoting from the opening pages of *When Prophecy Fails,* Festinger's account of that final night at Dorothy Martin's house:

Suppose an individual believes something with his whole heart; suppose further that he has a commitment to this belief, that he has taken irrevocable actions because of it; finally, suppose that he is presented with evidence, unequivocal and undeniable evidence, that his belief is wrong: what will happen?

Festinger and two colleagues asked Dorothy Martin if they could observe the Seekers as they waited. Festinger describes, moment by moment, what happened:

When the...clock on the mantel showed only one minute remaining before the saucer was due, [Dorothy Martin] exclaimed in a strained, high-pitched voice: "And not a plan has gone astray!" The clock chimed twelve, each stroke painfully clear in the expectant hush. The believers sat motionless.

One might have expected some visible reaction. Midnight had passed and nothing had happened...But there was little to see in the reactions of the people in that room. There was no talking, no sound. People sat stock still, their faces seemingly frozen and expressionless.

The Seekers stayed rooted in their seats for hours, slowly coming to terms with the fact that no visitor from outer space would be coming to their rescue. But did "disconfirmation" of their belief cause them all to abandon it? No. At 4:45 that morning, Martin announced that she had gotten *another* message. Because of the unwavering faith of the Seekers, she said, God had called off the destruction of the world.

What did Festinger make of all this? The more you invest in a set of beliefs—the greater the sacrifice you make in the service of that conviction—the more resistant you will be to evidence that suggests that you are mistaken. You don't give up. You double down.

As Festinger recalled in an oral history, "One of the things we expected would happen would be that, after the disconfirmation of this prediction...they would...have to discard their belief, but to the extent that they were committed to it, this would be difficult to do."

Back to the disaster of the Schweinfurt raids and the long discouraging summer and fall of 1943. Did those events lead Haywood Hansell and the Bomber Mafia to give up? Of course not. Here is what Hansell wrote to Ira Eaker after the first attack on Schweinfurt, on August 17: "I need not say how tremendously proud I was of the Regensburg-Schweinfurt operation. In spite of the very heavy losses, I believe it was completely justified and represents one of the turning points of the war."

Which is, of course, delusional. Schweinfurt was not a turning point of the war. But if you asked Hansell why he believed that, he would have given you his reasons. They were still learning. They got unlucky with the weather. They should have gone back the next week and hit it again, and then again, until every plant was completely destroyed.* Or maybe ball bearings weren't the best targets after all. But there were others, weren't there? What about oil refineries? That's how a true believer's mind works.

But outside that tight-knit circle was another man: Curtis LeMay. Like everyone else, he'd been to the Air Corps Tactical School, down at Maxwell, for his obligatory training. Yet he was never part of the Bomber Mafia circle. There was something in LeMay's makeup—in his obsession with the how and the what— that resisted any intellectual enthusiasms. He could make sure the pilots flew long and straight toward the target. He could instill in them the discipline not to bail out in panic along the way. He could train them to take off in fog. He was drawn to practical challenges. But doctrine left him cold.

In a 1971 interview, LeMay was even more blunt. He said he'd never been convinced by the elaborate logic behind the Schweinfurt raids: "The idea was, they found the ball-bearing plants over there—some of these swivel-chair target analysts back in the Pentagon—

---

* In his memoir, Hitler's minister of armaments and war production, Albert Speer, provides a detailed account of the Schweinfurt missions and what he calls "the enemy's error." He notes: "The attacks on the ball-bearing industry ceased abruptly. Thus, the Allies threw away success when it was already in their hands. Had they continued the attacks…with the same energy, we would quickly have been at our last gasp."

Carl L. Norden was a brilliant Dutch inventor who single-handedly invented the Norden bombsight used by the United States in World War II. Dubbed "the football" by airmen, it weighed fifty-five pounds and allowed bombardiers to factor in many variables, including altitude, wind speed, and airspeed. The legend goes, it enabled a bombardier to drop a bomb into a pickle barrel from six miles up.

The Bomber Mafia: Harold George (above left), Donald Wilson (above right), Ira Eaker (below left), and others were convinced that precision bombing, aimed at crucial choke points of the enemy's supply chain, could win wars entirely from the air. Their futuristic thinking was typical of what would become the Air Force Academy, whose modernistic chapel (below right) contrasts radically with the traditional architecture of the chapels at West Point and Annapolis.

Frederick Lindemann (at far left), who was Churchill's close adviser, believed that bombing should be used to break the will of the enemy by striking at cities indiscriminately. He is pictured here watching a display of anti-aircraft gunnery with Churchill (second from right) and other British military officials.

Royal Air Force marshal Arthur "Bomber" Harris ran the British bomber command using an "area-bombing" strategy, targeting military and civilian outposts alike.

The B-17 Flying Fortress, developed as a high-flying long-range bomber and used widely in the European theater, bombs an aircraft works in Germany.

A B-29 Superfortress waits to take off from a runway in the Pacific theater. The B-29 could fly faster, higher, and farther than any other bomber in the world and finally put the US Army Air Forces within striking distance of Japan.

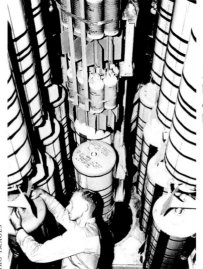

A crewman checks bombs in the cargo bay of a B-29 before the bombing of Tokyo.

Harvard chemistry professor Louis Fieser and his associate E. B. Hershberg (not pictured) conducted experiments with combustible gels that led to the invention of napalm.

The first napalm bomb test was conducted on July 4, 1942, behind Harvard Business School in Cambridge, Massachusetts.

To analyze the power of incendiary bombs, a perfect replica of a Japanese village was built at the Dugway Proving Ground, in Utah, in 1943.

In January of 1945, Major General Curtis E. LeMay (left) replaced Brigadier General Haywood Hansell Jr. (center) as head of the Twenty-First Bomber Command in the Mariana Islands. At right is Hansell's chief of staff, Brigadier General Roger M. Ramey.

The Army Air Forces's early facilities on Guam were primitive: tents and metal Quonset huts.

Aerial view of Tokyo bombing. On the night of March 9–10, 1945, one observer noted that the glow from the fires was visible 150 miles away.

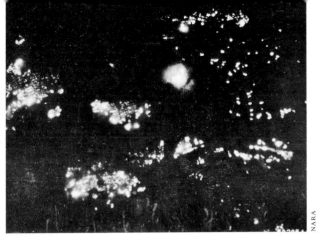

During Operation Meetinghouse, the single most destructive bombing raid of World War II, 1,665 tons of napalm were dropped on Tokyo, and as many as one hundred thousand people died.

General Curtis E. LeMay in 1954

The Center of the Tokyo Raids and War Damage is located in an unassuming building in Tokyo, Japan.

and the idea was, if we knock out that plant, which supposedly had the bulk of the ball-bearing production in the country, then the war would grind to a halt because there were no bearings."

Some swivel-chair target analysts back in the Pentagon. He's talking about Haywood Hansell and the Bomber Mafia, with their fanciful conjecture about how to disable the enemy.

LeMay continued, "The plan was okay—basically okay—but here we are trying to find something to win the war the easy way, and there ain't no such animal."

All that mattered to Curtis LeMay was the final outcome. He lost twenty-four planes on the decoy mission to hit Regensburg. Each of those bombers had a crew of ten, which meant that 240 men did not return to base. That's 240 letters that had to be written the next day by LeMay and his squadron leaders. *Dear Mr. and Mrs. Smith. Your son…Dear Mr. and Mrs. Jones. Your son*—240 times. And for what?

An Air Force officer named Ken Israel knew LeMay in the general's final years. They used to hunt together.[*] Once, Israel went to LeMay's house, in Southern California, to deliver some pheasants they had shot at Beale Air Force Base, just north of Sacramento. As Israel recalled:

---

[*] LeMay also had a shooting range in his basement. Naturally.

I rang the doorbell. He answered it, and he invited me to come in. I said, "Sir, I have your pheasants here." You walk into his foyer, and it was all marble. There on the wall to the left was a huge mural of Regensburg...On the opposite wall was a mural...a picture of Schweinfurt.

So I said, "Sir, is that Regensburg and Schweinfurt?" He said, "Yes, sonny." He just said, "Yep, we lost a lot of good men."

In the end, Curtis LeMay would have one of the most storied careers any Air Force officer would ever have. He planned or commanded countless missions more consequential than the Regensburg-Schweinfurt Raid. In 1948 and 1949, he would run the Berlin Airlift, one of the pivotal events at the start of the Cold War. He would eventually control America's nuclear arsenal as head of the Strategic Air Command. During his time in the service, he met every world leader imaginable, posed for pictures with the kinds of people the rest of us only read about in history books. He could have hung mementos of any of those things in his foyer. But he didn't. In the entryway to his house, he hung a reminder from his first real encounter with the orthodoxy of the Bomber Mafia, a reminder of failure and loss.

Part Two

The Temptation

Author's Note

Part Two of *The Bomber Mafia* takes place in Guam and Japan and all points east. But before we get to that, I want to tell a story from closer to the present.

I traveled to Tokyo when I was researching this book, along with my podcast producer, Jacob Smith. And right after we landed, Jacob and I got in a cab and went to visit a museum called the Center of the Tokyo Raids and War Damage. It's a memorial to the events that I'm going to describe in the next few chapters — the outcome of the struggle between the Bomber Mafia and Curtis LeMay.

I go to war museums all the time, such as the Imperial War Museums in London. The one on Lambeth Road is in a big grand building, but there are also two other branches in London and two more around the country. You can spend a few weeks going through them. And memorials. I've been to many of those, too: the Vietnam Veterans Memorial, on the Mall in Washington, DC; Yad Vashem, in Jerusalem. Each is powerful, moving, designed by a world-famous architect. Each has a *presence*.

So when Jacob and I got in our taxi in Tokyo, I assumed that we would be going toward the area where the museums are—the center of town, near the Imperial Palace. But we didn't. We went in the opposite direction, away from the business districts and tourists. We went east, down a very plain commercial street, over a big bridge. Farther and farther. Then we took a left-hand turn down a side street, and the driver stopped. And I wondered—was there some misunderstanding? I'd written down the address on a piece of paper. Did I write it down wrong? I showed the address to the driver. He nodded and pointed. And sure enough, when I squinted, I could see the sign for the museum. We were in front of what looked like a medical office building. It was three stories tall, built of brick.

We walked in and saw a little gift shop to the side—actually, just a couple of bookshelves. Next to that was what looked like a classroom, with a bunch of folding chairs, where an introductory video was playing. Then we went through a tiny courtyard and up the stairs to the main exhibition. The floors were linoleum. There were lots of black-and-white photos on the walls. A scale model of a B-29—the kind you'd buy in a toy store—hung from the ceiling. Jacob took a picture of me in front of the museum after we were finished. I have it on my phone. It looks like I'm coming from a dentist's appointment.

We're all familiar with the two atomic bombs that were dropped on Hiroshima and Nagasaki in August of 1945: Little Boy and Fat Man, dropped from the *Enola Gay*. There are grand monuments and memorials to *those* events. There are rows upon rows of history books that cover the topic. Debates continue to this day. I was in the midst of finishing this book when the seventy-fifth anniversary of those attacks was observed; on that day you had a hundred chances to relive the memory.

But the Center of the Tokyo Raids and War Damage is not about what happened after the nuclear attacks on Japan. It's about what happened before them—between November of 1944 and the late winter of 1945. From the command of Haywood Hansell to that of Curtis LeMay. A little bit of history that has been relegated to a side street.

Why is it on a side street? In some sense that's the subtext of the second half of this book. Something happened when the Bomber Mafia and Curtis LeMay moved their focus to the other side of the world, from England and Europe to the Mariana Islands, in the middle of the Pacific, something that everyone involved found inconvenient. Or unbearable. Or unspeakable. Or maybe all three.

This is not a war story but rather a story set in war, because sometimes our normal mechanisms of com-memoration fail us. And what comes next is an attempt to figure out why.

CHAPTER SIX

"It would be suicide, boys, suicide."

1.

All war is absurd. For thousands of years, human beings have chosen to settle their differences by obliterating one another. And when we are *not* obliterating one another, we spend an enormous amount of time and attention coming up with better ways to obliterate one another the *next* time around. It's all a little strange, if you think about it.

Nonetheless, even within that general category of absurd, there is a continuum. The war that was fought in Europe at least *resembled* previous wars. It was absurd in a familiar way: neighbor against neighbor. The D-day landing required a short trip across the English

Channel. People can *swim* the English Channel. On the ground, troops marched, holding rifles. They fired big pieces of artillery. Give Napoleon one week of training, and he probably could have managed the Allied push across Europe as well as any general from the twentieth century.

But the Pacific theater? It was on the other end of the war-absurdity continuum.

The United States and Japan probably had less contact with each other and knew less about each other than any two wartime combatants in history. More importantly, they were as far apart geographically as any two combatants in history. The Pacific war was, by definition, a sea war—and, as the conflict grew more intense, an air war. But the sheer scale of the Pacific battleground made it the kind of air war that no one had fought before.

For example, at the time of the attack on Pearl Harbor, the workhorse of the US Army Air Forces was the B-17 bomber, also known as the Flying Fortress. That's what LeMay and Ira Eaker and Hansell were using in Europe. The Flying Fortress had a range of roughly two thousand miles—one thousand miles out and one thousand miles back. In January of 1944, you couldn't find an air base controlled by the Allies within a thousand miles of Tokyo. Australia is more than four thousand miles from Japan. Hawaii is just as far. The Philippines made the most sense on paper, but the Philippines had been captured by the Japanese and weren't fully recaptured until late in 1945. In any case, Manila was still 1,800 miles from Tokyo.

If you were the United States and you wanted to drop bombs on Japan, how would you do it? Solving that problem took the better part of the war. The first step was building the B-29 Superfortress, the greatest bomber ever built, with an effective range of more than three thousand miles.

The next step was capturing a string of three tiny islands in the middle of the western Pacific: Saipan, Tinian, and Guam. They were the Mariana Islands, controlled by the Japanese. The Marianas were 1,500 miles across the water from Tokyo—the closest possible spot where you could build a runway. If you could put a fleet of B-29s on the Marianas, you could bomb Japan. The Japanese knew that, too, which led to another absurd moment: some of the ugliest fighting in the entire war was over three tiny clumps of volcanic rock that no one outside the western Pacific—no one— had so much as heard of before the war started.

The Marines were called in. One veteran, Corporal Melvin Dalton, recalled the fight:

> Our main job was to soften them up so the troops in the landing barges could get on the beach.
>
> After two or three days of that…the next morning at the crack of dawn, the ocean was full

125

of ships and barges headed for the beach, and there was gunfire you just can't believe. [*tears up*] The dead bodies were everywhere, just floating. Nobody had time to pick them up. They were all picked up later. When those Marines hit those beachheads, it was terrible sometimes.

One by one, over the summer of 1944, the islands fell to the US Marines,* whereupon Haywood Hansell was dispatched from Washington to head up the newly formed Twenty-First Bomber Command. It was an elite force composed entirely of the newest and most lethal weapon in the Air Force's lineup, the B-29 Superfortress. Its task was to cripple the Japanese war machine from the air, to pave the way for what the military leadership considered inevitable: a land invasion of Japan.

Leading the air attack on Japan was the most important job of Hansell's career. At that point, it was probably the most important job in the entire Army Air Forces. But the air attack plan was—in every sense of the word—absurd. Deeply absurd. First, consider

---

* While the exact death toll remains unknown, it's estimated that more than fourteen thousand Americans were killed, wounded, or listed as missing in action by the end of the Marianas campaign. Nearly all the Japanese forces stationed on the islands, around thirty thousand men, were wiped out. Today, 5,204 names are inscribed on a memorial on the island of Saipan, overlooking Tanapag Harbor.

the B-29. In 1944, it was a brand-new airplane, rushed into service. It broke down. Engines caught fire. No one had been properly trained to fly it. It had all kinds of idiosyncrasies.[*]

And this new weapon was to be launched from just about the least hospitable place imaginable for an air force base. The Marianas are hot and humid, blanketed with mosquitoes. They suffer torrential rains. There were no proper buildings, or hangars, or maintenance facilities, or roads, just Quonset huts and tents.[†] Haywood Hansell—a decorated general, the man who wrote the air-war plan used against Hitler in Europe—was camping out like a Boy Scout.

Vivian Slawinski, a second lieutenant in the Army Nurse Corps, recalled what it was like on the island of Tinian in those early months after the United States took over. "It was a lot of rocks…And there were rats in the place. They were up in the rafters. That was one thing that I couldn't stand. They'd come down and

---

[*] One problem with the earliest versions of the Superfortress was that the engines easily overheated. If you were a B-29 pilot in those days, your biggest worry was the enemy shooting at you. Your second-biggest worry was that your engines would catch fire.

[†] Needless to say, when LeMay arrived, he remained impervious to these less-than-ideal conditions. In fact, he described the dismal features of the island to his wife with almost comical optimism: "The beach here isn't too bad. Not much coral and what there is [is] mostly rotten, so you don't get cut up on it. There are quite a few sea slugs around, but they don't bother you. This just blew off on the floor, so you will see some of the same red dirt that we had in Hawaii."

nibble at some people's hair. And a couple times they came up close to my hands…We didn't have a hospital. All we had were these Quonset huts."

When her interviewer noted that those huts were metal and must have been hot, she replied, "Oh, honey, we were hot everywhere."

The sole thing the Marianas had going for them was that they were within range of Japan. But even that was an exaggeration. The truth is that they were within range *only* under perfect conditions. To reach Japan, a B-29 first needed to be loaded up with twenty thousand pounds of extra fuel. And because that made the plane dangerously overweight, each B-29 also needed a ferocious wind to lift it off the runway. This was as crazy a situation as anyone faced throughout the whole war.

It gets worse. By late fall of 1944, Hansell was ready to launch his first major attack on Tokyo. He described it after the war to a class at the Air Force Academy, in Colorado Springs: "The first operation against Japan was called San Antonio One. It was coordinated with the Joint Chiefs of Staff strategy, which made the timing extremely important."

Hansell's fleet would launch on November 17, 1944. Everything was ready. The weather looked good. The Army set up the media—with flashbulbs, cameras, and microphones—along the runways at dawn. Hansell conducted the pre-mission briefing himself.

"Stick together. Don't let fighter attacks break up the formations. *And put the bombs on the target.*"

The planes lined up. They were weighed down with all that extra fuel for the return trip, about to take off with the help of the usual strong wind blowing down the runway.

Except, that morning, there was no wind.

As Hansell recalled, "The orders were out, the airplanes were warmed up, and they taxied out to the end of the one strip that we had, and at that time, the wind, which had been blowing constantly down the runway for the last six weeks, died down to nothing."

So Hansell's overloaded B-29s couldn't take off. Then the wind started up again, only in the opposite direction. Could he turn his planes around—all 119 of them—and still make his window for the mission? He couldn't. All he had was a single runway, only half paved. He had to scuttle the mission.

It got crazier. The weather changed a *third* time. Hansell continued,

And three or four hours later, we were in the midst of an intense tropical storm, a hurricane, a typhoon. It lasted about six days, left the camp just a quagmire. And in the meantime, the B-29s were all loaded with bombs standing by, the orders were out. We were very seriously worried for fear that there'd be a security leak. It was pretty much too late to change. I kept thinking every day, maybe we'll make it. We sent weather airplanes out through this hurricane to trace it up the coast; it went right on up our route to Japan.

As a result, it was…[a week] later, before we were able to get that mission off.

Hansell made these remarks to a room full of Air Force cadets in 1967. Most of his audience was headed to Vietnam—by the way, another war on the truly absurd end of the absurd continuum—so they were hanging on to Hansell's every word. He'd fought in Asia, the part of the world where they were likely to be going next.

Then someone asked the old general: Suppose the wind hadn't died down and then changed course? Suppose you *had* managed to launch all your B-29s that morning of November 17, 1944? As the cadet pointed out, "You would have lost your whole organization if you had gotten off in time."

Hansell replied: "Certainly would."

Hansell and the rest of the Army Air Forces had none of the sophisticated navigation electronics that exist today. His entire fleet would have been up in the sky. One hundred and nineteen B-29s, each with a crew of eleven. That's 1,309 men circling around and around, looking helplessly for a speck of runway lights in the midst of a typhoon while the needles on their fuel gauges hovered over empty. And then, one by one, they would have been swallowed up by the ocean.

The storm lasted for six days. Hansell continued, "A couple of hours earlier, a couple of hours' difference in this weather situation, would have lost the entire bomber command. Because there was no other place to go."

Haywood Hansell's faith in the doctrine of precision bombing had been tested once, in the disaster over Schweinfurt. And his faith had survived intact. On the Marianas, his conviction would be tested a second time, only this time by something that had never crossed the minds of the Bomber Mafia, back in the seminar rooms of Maxwell Field.

2.

At the same time in 1944 that Haywood Hansell was deployed to the Marianas, Curtis LeMay was also transferred from Europe to the Pacific theater, to head another newly formed elite bomber group of B-29s: the Twentieth Bomber Command, stationed in eastern India, near Kolkata (formerly Calcutta).

Kolkata is—as the crow flies—the Indian city closest to Japan. It's in the far northeastern corner of the country. And since British India was a safe haven, the idea was that the B-29s would take off from there, then fly to an airfield carved out of some pretty dodgy territory in China, near Chengdu. There, they would refuel, then fly on to Japan, drop bombs, come back to Chengdu, refuel, and fly home to Kolkata. Distance-wise, that's like flying from Los Angeles to Newfoundland with a refueling stop in Chicago.

And then the crucial fact: Between Kolkata and Chengdu are the Himalayas, the tallest mountain range in the world. The pilots called the Himalayas "the Hump." If you thought that an air war launched from the Marianas was absurd, well, this was much, much worse.

This is how LeMay described flying the Hump. And LeMay never complained about anything.

It was a grueling hell... The mountains were a veritable smorgasbord of meteorological treachery— violent downdrafts, high winds and sudden snowstorms—all served up in temperatures 20 degrees below zero. As if they needed any reminding, the crews could frequently glimpse the 29,028-foot peak of Mount Everest thrusting up through the clouds just 150 miles from their flight path.

Over the course of the war, how many American planes do you think crashed while trying to navigate over the Hump? Seven hundred. The flying route was called "the aluminum trail" because of all the debris scattered over the mountains.

It gets worse. The air base in Chengdu didn't have any aviation fuel. It was in the middle of nowhere— just a landing strip. Much later, one of LeMay's airmen, David Braden, recorded an interview with a former brigadier general in the Air Force, Alfred Hurley. Every pilot who flew the Hump complained about it:

> **Braden:** That was a crazy thing. The only way they could get gasoline to Chengdu was by flying the Hump. Sometimes, if they had a headwind, it took twelve gallons of a B-29's gasoline to bring one gallon over the Hump.
>
> **Hurley:** It was extraordinary.
>
> **Braden:** It was insane.

Then, even from Chengdu, most of Japan's territory was still beyond the B-29's range. The planes couldn't get as far as Tokyo and make it back. So the best they could do was nibble at the closest corner of Japan's southwestern tip, where there was only one factory worth the Allies' attention.

Braden recalled, "When they started flying out of Chengdu, they could reach Kyushu [Japan], but there was really only one target on Kyushu, and that was an iron- and steelworks...They flew a mission there, and everybody was just exhausted."

To give you an example of what LeMay faced, here's a typical mission, launched out of Kolkata, on June 13, 1944. Ninety-two B-29s took off from India. Twelve turned back before crossing the Hump. One crashed. So that's seventy-nine that made it to China. They refueled, took off again. One crashed immediately after takeoff. Four more turned back because of mechanical problems. Six had to jettison their bombs. One got shot down on the way to Japan. Then the weather was terrible over Kyushu, so only forty-seven actually made it to the steelworks, and of those only fifteen could actually see the target. By the time the mission was complete, they'd lost seven planes and fifty-five men. And a total of one bomb actually hit the target. One.

You send ninety-two B-29s halfway around the world, and all you get is one bomb on the target.

The Japanese had a field day with the Twentieth Bomber Command. As their most famous propagandist, Tokyo Rose, broadcast to Allied airmen: "Listen to me, boys: fly back over the Hump to India. I hate to think of all of you getting killed. We have too many fighter planes and too many antiaircraft for you to get through. It would be suicide, boys, suicide."

That was how the air war in the Pacific was going in the fall of 1944. Whose position was more absurd: Curtis LeMay's or Haywood Hansell's? That's easy. Guam to Japan was hard. But India to Japan was insane.

The better question, though, is what *effect* each man's absurd predicament has on his way of thinking. Let's start with LeMay, someone whose entire identity is about problem solving. It's how he made sense of the world. He's not a man of great personal charm and charisma. He's not some towering intellectual. He's a doer. As he put it much later: "I'd rather have somebody who is real stupid but did something—even if it's wrong he did something—than have somebody who'd vacillate and do nothing."

That's what LeMay values. So imagine that he is stationed in India, thousands of miles from the action, and he's being asked to solve a problem that *cannot* be solved. You cannot wage an air war with any effectiveness when you spend twelve gallons of aviation fuel getting over the Himalayas in order to deliver one gallon to the other side.

No amount of human ingenuity or single-mindedness could overcome the obstacle of the Himalayas.

In the many considerations and reconsiderations of LeMay's legacy, there have been all manner of theories about his motivation for what he would do the following spring, when he took control of the air war in the Pacific. I wonder if the first and simplest explanation isn't just this: when a problem solver is finally free to act, he will let nothing stand in his way.

Then there's Haywood Hansell. His predicament was different; he was the true believer.

3.

Haywood Hansell's first act when he arrived in the Marianas was to ask, as any upstanding member of the Bomber Mafia would, What is the critical vulnerability of the Japanese war economy? What should my new B-29s attack? The answer to him was obvious: the Japanese aircraft manufacturing plants. But where are the Japanese manufacturing plants?

As Hansell recalled, "We were on Saipan with about forty or fifty B-29s [and] a deadline of the thirtieth of October. We had a deadline for an operation against the Japanese aircraft industry…and we had no target folders; we didn't know where the Japanese aircraft industry was."

So a crew flew out from the United States in a B-29 that had been modified for aerial reconnaissance. They took hundreds of photos, which showed that the Japanese aircraft industry—in particular, the Nakajima Aircraft Company, known as Subaru today—was heavily concentrated in and around Tokyo. The Allies knew that Nakajima was responsible for a large share of all Japanese combat-aircraft engines. Hansell said, *Let's start by hitting that factory, and we'll cripple the Japanese fighting force.*

San Antonio One was that first crucial mission, the one that narrowly avoided being lost to a typhoon. After a week of waiting, Hansell's planes finally took off.

The B-29s took off from the Marianas, skimming over the ocean at several thousand feet. As they approached Japan, they climbed high in the air, out of harm's way. They turned at Mount Fuji, then came in from the west over Tokyo. Here, over aerial shots of the city, in the Army Air Forces's war film, Ronald Reagan describes what happened:

Six hours later, through the clouds, they saw it— Fujiyama [Mount Fuji], ancient symbol of Japan. Here come some modern symbols. Phosphorus bombs and flak. And fighters…Within a radius of fifteen miles of the Imperial Palace live seven million Japanese, a people we used to think of as small, dainty, polite, concerning themselves only with floral arrangements and rock gardens and the cultivation of silkworms. But it isn't silkworms and it isn't Imperial Palaces these men are looking for. In the suburbs of Tokyo is the huge Nakajima aircraft plant. Well, Bud, what are you waiting for?

He lays it on a little thick.

San Antonio One was hugely symbolic. It demonstrated that Japan could finally be reached. But was it a success, as a military operation? After the war, speaking to cadets at the Air Force Academy, Hansell tried to put a good face on things. "The operation wasn't as good as we would have liked, but as an initial effort, it did show it could be done. This was a very doubtful issue at the time."

The operation wasn't as good as we would have liked was, to say the least, an understatement. The first raid damaged a mere 1 percent of the Nakajima plant. Hansell tried again three days later. None of the bombs actually hit the plant. On December 27, he sent back seventy-two B-29s. They missed the plant but wound up setting fire to a hospital. In the end, Hansell went after that factory five times and barely touched it.

Part of the difficulty was the same problem the Bomber Mafia had had over Europe: clouds. The bombardiers looked for the target through their Nordens and couldn't find it. But there was another problem with the weather, a problem much worse and much bigger than anyone at the time could understand.

One of Haywood Hansell's B-29 pilots, Lieutenant Ed Hiatt, was later interviewed for a documentary by the BBC. He described one mission:

> After flying six hours, we climbed up to bombing altitude…We climbed up to thirty-seven thousand feet, and just as we broke out of the storm, there's Mount Fuji, sitting right in front of us. And it's a gorgeous sight, it really is.

Hiatt's bombardier, a man named Glenn, started to make his calculations on their Norden bombsight, focusing on the Nakajima factory. But the telescope on the bombsight wouldn't line up with the approaching target. Hiatt continued:

> He turned around, and he said, "I can't get this damn telescope on the target"…And so we called the radar operator to check our ground speed and…he came back and he says we've got a 125-knot tailwind. He said we're going about 480 miles an hour. It's impossible—it can't be. There's no winds like that.

There's no winds like that. No Army Air Forces pilots had ever experienced what was happening to the B-29 bombers over Japan. They never expected winds like *that.*

"We're going 480 miles an hour when we should be going 340 miles an hour…I said, 'Well, Glenn, drop the damn bombs.' He dropped the bombs, and we were already twelve miles past the target because of that wind," Hiatt said.

They were bewildered. And back at base, they couldn't explain it to their superiors.

> When they debriefed us, they gave us the third degree. They wouldn't believe us. "There's no such thing as a 140-mile-an-hour wind up there over Japan," they said. "No, there is no such thing. There can't be a wind like that. You're lying. You

didn't make it over the target; you're just making this up." And...we had our operations officer as a passenger with us, and he vouched for it. He said, "There was a wind that high."

The Twenty-First Bomber Command had a team of meteorologists attached to it. They'd been trained at the University of Chicago. Meteorologists were crucial to the success of bombing campaigns, particularly in the days before sophisticated radar. You had to know whether there were clouds over your target. Or whether there was a typhoon poised to swallow up your command.

But the tools available to meteorologists of that era were crude. I know this is a digression, but the easiest thing to forget about the Second World War is that it took place in another technological era. It's half twentieth century and half nineteenth century. The chief tool meteorologists had at that time were balloons, weather balloons that would float up into the atmosphere carrying little instrument kits that could record the wind, the temperature, and the humidity and transmit that information back to earth by radio.*

---

* Weather balloons are still used by meteorologists today. Twice a day, hydrogen- or helium-filled balloons are released simultaneously from around nine hundred locations worldwide. An instrument attached to the balloon, called a radiosonde, measures atmospheric pressure, temperature, and humidity and transmits the information back to tracking equipment on the ground.

John M. Lewis, a researcher at the National Severe Storms Laboratory, part of the Desert Research Institute, in Nevada, knew a number of the meteorologists who worked with the Army Air Forces during the war. I asked him if the weather balloons were connected back to earth with a rope. His reply: "Oh, no. They're released. They'll eventually, as the pressure gets lower as the balloon goes higher in the atmosphere—they expand, expand, expand. Kaboom! They explode, and they fall to the ground with the instrument attached. And at that time, they had a message on all the instrument packages: 'Could you please return this to the University of Chicago? Here's the address.'"

In the Pacific theater of war, that obviously wasn't going to happen.

So there they are, the meteorologists, in the middle of the Pacific, with one of the most important jobs in the whole outfit—figuring out when to send the bombers—and they're baffled. What's going on with these super-fast winds the pilots are reporting high over Japan?

I asked Lewis if they had any reason to suspect that the winds around Mount Fuji would be so incredibly high. His reply: "They did not reach their conclusions until the pilots came back."

After each bombing mission over Japan in 1944, the crews returned to the base and told the same story. As Ed Hiatt later recalled,

To tell you how powerful these winds were: a reconnaissance plane went up one time to take some pictures after a mission to see how effective they'd been, and the navigator called the pilot and told him they were going three miles an hour backwards. That was something you couldn't afford to do because if you went from east to west, you were gonna be a sitting duck for Japanese fighters or their flak.

The pilots had encountered what would come to be known as the jet stream, a river of fast-flowing air that circles the globe in the upper atmosphere, starting at around twenty thousand feet. A Japanese scientist named Wasaburo Ooishi had actually discovered the jet stream in the 1920s in a series of groundbreaking experiments. But Ooishi happened to be devoted to the artificially constructed language called Esperanto, which was briefly in vogue in that era, and he only published his findings in Esperanto, which meant of course that almost no one read them. And since almost no one had ever flown at the altitudes the B-29 was flying at, there were no firsthand reports of the jet stream winds, either. It was a mystery.[*]

---

[*] A few others encountered the jet stream after Ooishi. In the 1930s, a Swedish meteorologist named Carl-Gustaf Rossby identified and characterized both the jet stream and the type of atmospheric waves that would later be named Rossby waves. In 1935, the American pilot Wiley Post became the first to experience the jet stream directly.

As John Lewis explained it to me, "This fast stream of air, very narrow, moves from north to south in both hemispheres. Basically, it is dividing the very cold air of the polar regions from the more warm midlatitude and equatorial air."

When I asked him how wide the jet stream is, he replied, "I would say typically two hundred kilometers across, something on that order, certainly not a thousand kilometers, rarely five hundred kilometers, sometimes a hundred kilometers."

It was such a new discovery that nobody realized it circled the entire planet. Lewis explained, "That was not discovered until the early 1950s, when we started to make upper-air observations routinely over the United States [and] some of the countries in Europe."

The jet stream circles the whole earth, a narrow band of incredibly fast wind. It retreats to the poles in the summer and moves toward the equator during the winter months.

And in the winter of 1944 and early spring of 1945, this narrow, hurricane-force band of air was directly over Japan. That made it impossible for Hansell's pilots to do any of the precision bombing they had planned

---

Post was famous for his daring flight experiments and discovered the strong winds of the jet stream during one of his high-altitude transcontinental flight attempts. The term *jet stream* wasn't coined until a German meteorologist described the strong winds as *strahlströmung*, which translates literally to "jet stream."

to do. If they flew across it, the plane would get blown sideways. If they flew into it, they'd be fighting to stay aloft and would be easy targets for the Japanese. And if they flew with it, they'd be racing too fast to take proper aim.

The dream hatched back at Maxwell Field in the 1930s and brought to life by the genius of Carl Norden had run up against an unstoppable force in the skies over Japan.

This is not the same kind of obstacle as the Bomber Mafia faced over Schweinfurt and Regensburg. There, Hansell could justify to himself that the problem was solvable, that the first raid was a learning experience, that the raids could get better and more accurate. Every revolutionary understands that the path to radical transformation is never smooth. Software programmers have a beta version, and then a 1.0 and then a 2.0, because they realize that they can never get it right the first time.

But in the case of the jet stream over Japan, there was no 2.0 version, no revision that Hansell could use to bolster his faith. High-altitude precision bombing in the midst of a jet stream is impossible.

The dreams of revolutionaries go awry when they are forced to confront an unanticipated obstacle—not a rational obstacle such as inexperience or haste or miscalculation, but something immovable. And in that moment of vulnerability and frustration, with his dream in pieces all around him, Haywood Hansell, like Jesus in the wilderness, was presented with a temptation. As it says in the Bible:

> And Jesus, full of the Holy Spirit, returned from the Jordan and was led by the Spirit in the wilderness for forty days, being tempted by the devil.

And what did the devil do? He led Jesus to the top of a high mountain—in legend, the peak on the road between Jerusalem and Jericho—and offered him power over everything he could see.

> And the devil took him up and showed him all the kingdoms of the world in a moment of time, and said to him, "To you I will give all this authority and their glory, for it has been delivered to me, and I give it to whom I will. If you, then, will worship me, it will all be yours."

You can have everything. Victory over your enemies. Dominion over all you can see from twenty thousand feet. All you have to do is walk away from your faith.

CHAPTER SEVEN

"If you, then, will worship me, it will all be yours."

1.

Haywood Hansell's temptation requires a detour, just for this chapter, away from airplanes and bombing runs and high winds over Japan to a meeting. A secret meeting, early in the war, in Cambridge, Massachusetts.

The president of MIT was there, along with, among others, a Nobel Prize winner, the president of the Standard Oil Development Company, and two professors—Louis Fieser of Harvard and Hoyt Hottel from MIT, a giant in his field who would later become the group's chairman and spiritual leader.

The meeting was held at the behest of what would become the National Defense Research Committee.

The NDRC was the government group charged with developing new weapons for the American military. Its most famous effort was, of course, the Manhattan Project, the multibillion-dollar operation out of Los Alamos to develop the atomic bomb. But the scale of the war effort was such that the NDRC had many other projects under way as well. It had Americans, off in corners, working on schemes shrouded in darkness. Missions launched that no one heard about. Ideas being pursued in one place that contradicted ideas being pursued in another place. During the war years, to use the cliché, the right hand of the United States government did not always know what the left hand was doing. And one of those shadowy left-handed projects was Hoyt Hottel's subcommittee.

Unlike the geniuses down at Los Alamos, the men weren't physicists. Their job was not to find better ways to blow things up. They were chemists. Specialists in the particular consequences of combining oxygen, fuel, and heat. Their job was to find better ways to burn things down.

As Hoyt Hottel recalled after the war, "Come '39, a lot of people thought that a war was something we'd be in sooner or later, and our state of preparedness was poor... We needed to know more about incendiary bombs."

Hottel's group of chemists and industry officials and Nobelists began to meet whenever they could. They planned; they tinkered; they schemed. And on May 28, 1941, at a session in Chicago, they had their first real breakthrough. Hottel told his committee about a strange incident that had just happened at a DuPont chemical plant in Delaware. A group there had been working with something called divinylacetylene. It's a hydrocarbon—an oil by-product—and if you mix it with a pigment, the paint will dry into a tough, thick adhesive film. But the film kept bursting into flames, which was a problem for a paint company such as DuPont. For the fire obsessives on the NDRC chemistry committee, however, that was *fascinating*.

Around the table, one man raised his hand. *I'll look into that.* It was the Harvard chemistry professor, Louis Fieser.

Fieser was born in Ohio in 1899. He majored in chemistry at Williams College, got his PhD from Harvard, and earned postdoc fellowships at Oxford and Frankfurt. Before the war, he was the first to synthesize vitamin K. His research assistant was his wife, the equally brilliant Mary Fieser. Women didn't get hired as chemistry professors in those days, but together, the couple wrote one of the definitive chemistry textbooks of the twentieth century. Louis was largely bald and a little heavyset. He sported a mustache and was always with a cigarette.

Louis Fieser was also a man of imagination and whimsy. His scientific memoir, published in 1964, begins with his wartime work, but then quickly turns to detailed descriptions of things such as a pocket firebomb that he called, in an inspired bit of brand awareness, the Harvard Candle. There is a chapter about attaching incendiary devices to bats. There is an extended riff on how to ignite a thousand-gallon oil slick. Detailed plans for a squirrel-proof bird feeder. And, the coup de grâce, a chapter about one of his many cats, a Siamese called Syn Kai Pooh.

In the Science History Institute archives, there's an extended interview with a colleague of Fieser's named William von Eggers Doering, who taught chemistry for years at Yale and Harvard. The interview goes on for hours—and it's weirdly riveting. It gives you a glimpse into a world of scientists who had license to be just a little mad. This is how Doering remembers working in Fieser's laboratory at the very beginning of the war:

God, what was the compound we were after? Oh, yes, trinitrobenzyl nitrate [*laughter*]...Listen to this: you put it—do you remember those heavy Carius tubes? They were for some sort of an analysis where you digested something with nitric acid at high temperature. So these were eighth-inch-thick tubes, about an inch in diameter and a couple of feet long. So you put in about twenty or thirty grams of TNT, you poured [in] a little excess of bromine, no solvent. You sealed the

damn tube, put it in a bomb—an iron bomb—
you know, with a wire wrapped around it to raise
the temperature [*laughter*]…So that in effect, if
you put the heating tube in that little space, then
if it blew up, the glass would hit this little part
of the wall [*laughter*] on the left and the other on
the right. Well, of course, half the tubes blew up!
[*laughter*]

Understand that Doering was one of the great chemists of his generation. He published his first scientific
paper in 1939 and his last in 2008—eight decades of
work. In every picture I've seen of him, he's wearing
a polka-dot bow tie. But in this interview, he's like a
thirteen-year-old kid with a chemistry set:

The laboratory would be filthy with bromine,
and you wondered when the TNT was going to
detonate! [*laughter*]…Oh, God, it was marvelous
times! The Germans have a word to describe certain persons as *tierisch ernst*, which means having
an animal-like seriousness about them. I must say
there was very little of that [*laughter*] in those
days! [*laughter*]

When Louis Fieser came down to the lab, smoking
his ever-present cigarette, the grad students would play
pranks on him.

Louis would come in to talk to his people and would invariably throw his cigarettes, still burning, into the sink. And so the game was to try to guess when he was coming down and then pour ether in [*laughter*] the sink in the hope that it would catch fire. [*laughter*]

In the hope *that it would catch fire!*

Fire was not just of intellectual interest to the people in Fieser's basement lab. It was also an obsession, a fixation. So when Hoyt Hottel told the subcommittee that something in one of DuPont's paint mixes would spontaneously burst into flames, who instantly raised his hand? Fieser, of course. *I'll look into that.* And to help him with his investigation, Fieser immediately turned to another member of his basement coterie. In his memoirs, he writes, "I volunteered chiefly because I had available in my peacetime research group a man ideally qualified to experiment with and evaluate a hazardous chemical. Dr. E. B. Hershberg."

I spoke to E. B. Hershberg's son Robert Hershberg and asked him how his father first connected with Fieser. Robert replied: "First, he's from the Boston area, [and] I think the very quick and short answer was there were limited places for employment for Jews, and Fieser couldn't care less about religion. So that's the lab he wound up in."

E. B. Hershberg was, in Louis Fieser's words, "a masterful experimentalist in organic chemistry... also versed

in engineering, in mechanical drawing, in carpentry...
and in photography...Furthermore Hershberg...was
experienced in the handling of military explosives, fuses,
poison gases, smoke pots, and grenades" and had invented a long list of devices, including "the Hershberg
stirrer, the Hershberg stirring motor, and the Hershberg
melting point apparatus."

As Robert recalled:

> In our basement we had defused bombs and things
> of that nature, and [I have] pictures of explo-
> sions that occurred. And some of the incendiary
> devices were in the desk drawers...There were
> things like notebooks that had incendiary devices
> in them, that if you were captured, you pulled
> the pen out, [and] you had half an hour to write
> everything down and what you wanted and get
> out of there before it blew up and burned down
> the building.

That was E. B. Hershberg.

So Louis Fieser went to Delaware to investigate
the DuPont compound that made paint catch on fire:
divinylacetylene. After he returned to Harvard, he and
Hershberg started cooking up batches of it. They would
put the batches in pans and place them on the window-
sill of Fieser's basement lab. They noticed that the
substance gradually changed from a liquid to a thick,

viscous gel. They poked the gel with sticks. Then they set fire to it and noticed—and I'm quoting here from Fieser's book, because this was the crucial insight—"that when a viscous gel burns it does not become fluid but retains its viscous, sticky consistency. The experience suggested the idea of a bomb that would scatter large burning globs of sticky gel."

You drop the bomb, and the gel scatters. And it doesn't just burn itself out. Big globs of gel fly in every direction, and those globs stick to whatever surface they land on—and keep burning and burning and burning.

Hershberg and Fieser now had to find a way to test this new concept of incendiary gels. So they built a little two-foot-tall wooden structure in the lab and compared how well various gel formulations did in burning it down. Divinylacetylene was good. But a gel made of rubber and benzene was better. And gasoline was even better than benzene. They tried amber-colored smoked sheet rubber. Pale crepe rubber. Rubber latex. Vulcanized rubber. They made a prototype and took it with them in a suitcase on the train to Maryland, giving it to the porter to carry. The porter said, "It feels heavy enough to be a bomb."

Next they tried aluminum naphthenate, a sticky black tar made by a chemical company out of Elizabeth, New Jersey. The tar didn't mix well with gasoline, but they solved the problem by mixing in something else called aluminum palmitate. Gasoline mixed with aluminum *na*phthenate plus aluminum *palm*itate.

Napalm.

Robert Neer, author of *Napalm: An American Biography,* told me why napalm is so effective:

If you want an effective incendiary, something that is sticky is much more effective than something that is not sticky, because it actually adheres to whatever it is transferring its radiation energy into. And that's why napalm is so effective.

If the jelled material is too soft or too weak, then it won't actually deliver a very large amount of radiation to whatever it's sticking to. You can think of a Molotov cocktail that's filled up with gasoline, exploding and delivering gasoline. It can burn somebody or something quite terribly, but the fire will go out relatively quickly. Whereas by contrast, if napalm is thrown on something, it will stick to it.

A gel that was too loose would produce what they described dismissively as applesauce. In other words, it wasn't thick enough or solid enough in its globules to adhere to something. And something that was just right would form quite large-size chunks. It had to be a balance between too thick and too thin and just right. And that's what they ultimately hit upon with napalm.

Neer and I visited the Harvard soccer field, right behind the business school, which is across the river from the main campus. It's where Hershberg and Fieser tested napalm in 1942. Hershberg had figured out how to turn their new gel into a bomb: by inserting a stick of TNT with a layer of white phosphorus wrapped around it in the middle of a canister of napalm. Phosphorus burns at a very high temperature, so the TNT would go off, driving the burning phosphorus into the napalm gel, igniting it, and sending globs of it in every direction. For a bomb case, they used a shell that had originally been designed to hold mustard gas. Robert Neer described the scene:

It was on Independence Day, 1942. They had finalized the formulation for the gel incendiary on Valentine's Day, February 14. And then they figured out the white-phosphorus-burster ignition system and got the bomb shells from the military and built their prototypes.

They dug a lagoon into the field. The lagoon was, I believe, about a hundred feet in diameter. It was quite a substantial lagoon because they didn't want anybody to get hurt. And they had this pretty large napalm bomb in a canister that they were going to explode in the center. So they put the bomb right in the center of this lagoon, which had been filled up with water by some trucks from the Cambridge Fire Department.

The birth of napalm. Baptized in eight inches of water in the middle of Harvard's soccer pitch. When he was doing his research, Robert Neer spotted a little detail in the photos from that day.

In the initial pictures of the test, there are people dressed in whites playing on the tennis courts. And then after the bomb goes off, you see that the tennis courts are abandoned...So maybe they told everybody that they were about to test this napalm bomb, or maybe they just let them keep playing tennis and then tested it and everybody ran away. I don't know. Nobody was injured in these tests. After the bomb was exploded, they made a very careful catalog of the distribution and size of the extinguished globules of napalm, because that was part of determining the most effective consistency of the gel.

Fieser and Hershberg took their creation back to the National Defense Research Committee, and Hottel realized he had finally found what they were all looking for: napalm, created at Harvard University, perfected in the fields along the meandering Charles River.

2.

There was never any question what napalm was for. It was intended to be used against Japan.

A few months after Pearl Harbor, two American analysts published an essay in *Harper's Magazine*. When it comes time to retaliate against Japan, the authors argued, there's a really easy way to do it. Fire. Osaka was their case study. Osaka's streets are very narrow. Narrow streets means that fire can jump easily from one side to the other. And the city didn't have a lot of parks that could act as firebreaks.

Plus, unlike Western cities, Japanese cities weren't built of bricks and mortar. The beams, joists, and floorboards of houses were all wooden. Ceilings were made of heavy paper soaked in fish oil. Walls were made of wood or thin stucco. Inside were tatami—straw mats. Japanese houses were tinderboxes.

As the analysts wrote, "After some considerable calculation, we have determined that the combustible coverage in the twenty-five-square-mile area that is the central section of Osaka is 80 percent, as opposed to 15 percent for London."

Eighty percent—that's almost the whole city.

The people writing the article weren't military officers or White House policy makers. The idea that you might destroy 80 percent of one of your enemy's cities—burn it to the ground—was heretical. William Sherman, the general who led the Union Army on its final devastating course through the South after the Civil War, famously burned down Atlanta. But not all of Atlanta. The business and industrial districts. Not civilians in their homes. In the aftermath of Pearl Harbor, however, this heretical idea began to seem less heretical. Didn't a lot of Japanese industrial production actually take place in people's houses? Wasn't it true that a lot of the war effort happened in living rooms as well as factories? A gradual process of rationalization began to take hold.

Army War College historian Tami Biddle explains,

Regarding Japan, we still told ourselves, *Well, there's lots of industry in cities,* which is what the British had told themselves when they switched over to area bombing.

If you are a morally guided person, and you want to be able to sleep at night and reconcile what you're doing with your own principles, you've got to find language and concepts to tell yourself that what you're doing is okay…

The decision at that point was *Okay, gloves come off. We have to do whatever we can do to bring this nation down.*

Hoyt Hottel heard those whispers, those rationalizations. Did he read that *Harper's* essay? He must have. The NDRC told him to investigate the utility of incendiaries as weapons of war, and so he decided—good scientist that he was—to put this new weapon, napalm, to the test. He set up one of the most elaborate experiments of the war: an incendiary demonstration test at Dugway Proving Ground, the Army's eight-hundred-thousand-acre test facility in the middle of the Utah desert.

As Hottel recalled, "These generals don't believe what scientists do. They only believe what they think they can visualize. We've got to build a Japanese village and a German village. It's amazing the enormity of the effort that went into building those things." They built two sets of perfect replicas of enemy houses on the sands of the Utah desert.

Hottel brought in top-level architects. For the German village, he called on Erich Mendelsohn, a brilliant German Jewish architect who had designed some of the most beautiful art deco and art moderne buildings of the 1920s and 1930s. For the Japanese village, Hottel conscripted Antonin Raymond, who had lived in Japan for years and to this day is probably Japan's most celebrated Western-born architect.

Hottel recalled how much care went into the replica villages: "We decided that the two-inch-thick rice-straw mats that characterized the Japanese home, the tatami, were important because they were the major resistance to the bomb passage through one floor after another. So we had to have tatami."

They built twenty-four Japanese residences—twelve complexes with two units each. They included shoji—Japanese sliding screens—and perfect replicas of Japanese window shutters.

Antonin Raymond also set exacting standards. Hottel recalled, "Raymond wanted the cabinetwork on making these things under his eye in New Jersey. Here we wanted to build a place in Utah, the wood was in the Pacific, the cabinetwork was to be in New Jersey—and these are absurdities."

Hottel's project manager, Slim Myers, was another perfectionist. "Slim said, 'Damn it, we've got to be absolutely right. These generals are not going to stop us because we didn't have something that was really characteristic. We've got to be right.'"

By the summer of 1943, Hottel's model villages were ready for their tests. The military assigned a fleet of bombers to Dugway. One plane after another dropped its incendiaries. And after each round, the teams on the ground rebuilt whatever was damaged. Hottel first tried British thermite bombs, which were favored by the RAF commander Arthur Harris in his night raids on Germany. They compared those results with those of Hershberg and Fieser's napalm, packed inside bombs that went by the name M69. Hoyt Hottel and his team stood by, keeping score.

Hottel recalled, "We early [on] decided that we couldn't wait for the fire truck. We had to rush out to take care of fires. In fact, we had to rush out before all of the bombs had dropped."

Hottel grouped whatever fire he saw into three categories of destructiveness: (a) uncontrollable within six minutes, (b) destructive if unattended, and (c) nondestructive. Napalm was the hands-down winner, with a 68 percent success rate in the first category on Japanese houses. It caused uncontrollable fires. By contrast, British thermite ran a poor, distant second. With napalm, the United States had built itself a superweapon. And the Army was so proud of its new bomb that it made glowing promotional films about it.

> The main component of the M-69 bomb [is] a cheesecloth sock containing specially processed jellied gasoline. When ignited, the gel filling becomes a clinging, fiery mass, spreading more than a yard in diameter...It burns at approximately one thousand degrees Fahrenheit for eight to ten minutes...For air drops, the M69 is assembled in groups of thirty-eight...The cluster is released and opened, and the individual bombs, with gauze streamers trailing, drop toward the target.

3.

Imagine that you were a member of the Bomber Mafia and you happened to sit in on that demonstration test at the Dugway Proving Ground. You saw the meticulous reconstruction of Japanese villages. Heard the B-29s— *your* B-29s—screaming down the skies to drop their fiery payloads. You saw the houses engulfed in flames. What would you have made of it all?

I'm guessing you would have been baffled. The Bomber Mafia was consumed with the potential of the Norden bombsight, a machine that used technology to redefine war, to make it more humane, to restrain the murderous impulses of generals on the battlefield. If you weren't using human ingenuity and science to improve the way human beings conducted their ruinous affairs, then what was the point? This is what technological innovation was *for*.

But suddenly you were standing somewhere deep in the Utah desert, under a hard sun, observing a military exercise authorized and funded by the same US military that paid for your Norden bombsight. Except that these people are using science and ingenuity to create *incendiaries*, objects to be dropped from the sky with the intention of starting violent, indiscriminate fires. You had been going to elaborate pains to avoid hitting anything but the most crucial industrial targets. Now the Army was using your precision-bombing apparatus to obliterate people's houses. Here was the government—your own military bosses back in Washington—pursuing a strategy 100 percent in violation of your principles. And that's not even mentioning the top-secret work in the New Mexico desert, where the smartest people in the world were being given billions of dollars to create a weapon so devastating, so catastrophic in its effects, that it would change world politics forever. If firebombs were a betrayal of precision-bombing doctrine, then what was the atomic bomb? Good Lord. It was a technological Judas.

But then, after the initial outrage had passed, you might well have had a second thought. An unbidden thought. A temptation.

Because napalm would solve all the problems Haywood Hansell and all his precision bombers had had in the war thus far. Precision bombing wasn't working. Hansell was struggling under some of the most difficult conditions faced by any combat commander in the entire air war. His planes couldn't hit what they wanted to hit because of the high-altitude winds and the clouds over Tokyo. So maybe, the thinking went, don't bother *aiming* at anything at all. Just burn everything down. The place is a tinderbox. *All Haywood Hansell had to do was switch to napalm.* He could carry out morale bombing against the Japanese, only with a weapon far, far deadlier than the bombs the British used on Germany. *Sixty-eight percent success rate in category (a) on Japanese houses, where the fires became uncontrollable within six minutes.*

In the Bible, Jesus spends forty days and forty nights in the wilderness, being tempted by Satan. Haywood Hansell launched his first air strike on Japan on November 24, 1944. His last day as head of the Twentieth Bomber Command was January 19, 1945. That's fifty-five days in the wilderness of the Marianas when he was tempted to abandon all that he had fought for and believed in in exchange for the chance to defeat the Japanese enemy.

Over the course of those fifty-five days, the pressure on Hansell grew intense. The Army shipped thousands of napalm canisters to the Marianas. They urged Hansell to try—just try—a full-scale incendiary attack on Japan.

Hansell lost a B-29 on nearly every major mission. The margin of error for getting back to the Marianas was so slim that damaged planes would sometimes just plunge into the Pacific on the way home, never to be seen again. Morale dropped. The same General Hansell who had been almost absurdly upbeat about the prospects for precision bombing a year earlier now turned dark and angry. After yet another failed mission in which they missed the primary target entirely, one of Hansell's key officers, Emmett "Rosie" O'Donnell, held a briefing for his airmen. He was trying to keep their spirits up. "Boys. It's tough. It's a tough mission. But I'm proud of you, and we're doing well." Then Hansell stood up. And blasted the room.

"I don't agree with Rosie. I don't think you're earning your salt out here. And the mission. If it continues like it is...the operation will fail." Hansell embarrassed one of his officers in front of everyone, something no commanding officer should ever do, not if he wants to maintain the respect of his men.

Historian Stephen McFarland described Hansell to me this way:

> He's kind of a tragic character in a way. His forte was thinking. He helped formulate this strategy, helped design the war plans that would lead to the bombing of Germany and Japan. He was almost philosophical. He was more of a thinker. He was more of a—I don't want to say, pencil-neck-geek type of person.
>
> He was not a combat officer. He was not the great leader. He spoke in terms of high ideals...He never cussed, and commanders in the war who never cursed, they weren't much appreciated by the pilots. They wanted somebody who was down to earth, who understood what it was like.

By the end, Hansell was increasingly alone. Historian Tami Biddle put it like this:

> I think when a commander goes into a command with an idea of what's going to work, first of all,

165

they believe it. They have to believe it because you couldn't send so many men into combat if you just didn't believe in what you were doing.

You send men into combat with an idea, and you're anchored to that idea about what you've got to do to make it work and to justify those lives and to justify that blood and treasure…

I think commanders when they're in the field, doing something that's so intense as what Hansell was trying to do between basically October and December of 1944—he is fixated. I think he's got one thing on his brain, and he's just determined that he's going to make it work.

At one point, in late December, the second in command of the entire Army Air Forces, Lauris Norstad, gave Hansell a direct order: launch a napalm attack on the Japanese city of Nagoya as soon as possible. It was, in Norstad's words, "an urgent requirement for planning purposes." Hansell did a trial run and burned down a paltry three acres of the city. Then he grimaced, shrugged, delayed, promising to do something bigger at some point, maybe, when his other work was finished.

He wouldn't give in to temptation.

And because he wouldn't, Norstad flew in from Washington. You can imagine the moment. The visiting dignitary from home. An honor guard at the airfield. Whiskey, cigars, and gossip in Hansell's Quonset hut.

Then Norstad turned to Hansell, completely out of the blue, and said: *You're out. Curtis LeMay's taking over.*

"I thought the earth had fallen in—I was completely crushed." That's how Hansell later described his feelings in that moment. Hansell was given ten days to finish up. He walked around in a daze.*

On his last night in Guam, Hansell got drunk and sang for his men: "Old pilots never die, never die, they just fly-y-y away-y-y-y."

Curtis LeMay arrived for the changeover, flying himself to the island in a B-29. The two men posed for a picture together. LeMay said, "Where do you want me to stand?" The camera clicked.

After that, Hansell went home to run a training school in Arizona. His war was over.

"I got to read a number of interviews with the man," the historian Stephen McFarland told me. "I got to read a few of his letters, and he was truly a thoughtful, caring individual. And he was a true believer, but he was not the kind of man who was willing to kill hundreds of thousands of people. He just didn't have it. Didn't have it in his soul."

---

* Hansell's final mission takes place on January 19. It's a tremendous success. Sixty-two B-29s take out the Kawasaki factory. As historian William Ralph notes: "Every important building in the entire complex was hit. Production fell by 90 percent. Not a single B-29 was lost. Hansell flew back to the United States the next day." The irony is unbearable.

CHAPTER EIGHT

"It's all ashes. All that and that and that."

1.

Military historian Conrad Crane is an expert on Major General Curtis LeMay. I asked him about LeMay's mind-set when he became head of the Twenty-First Bomber Command after taking over from Haywood Hansell in January of 1945.

As Crane put it, "When he takes over the Twenty-First Bomber Command, when he first arrives in the Marianas, he does not have his eventual strategy worked out. His mind is still open." If Hansell was inflexible, a man of principle, LeMay was the opposite.

First things first. LeMay was not happy with the military's infrastructure on the Marianas. It was all built by the Navy's construction battalion, the Seabees. LeMay had lost none of his disdain for the Navy, the military branch he believed cheated in the bombing exercise years before.

As Crane related,

He looks around and sees the primitive nature of the facilities and said, "This won't do"... He gets invited to have a dinner with Admiral Nimitz, who also is headquartered in the Marianas, and he goes over to Nimitz's place and he's in this ornate... almost a palace, and he gets fed [a] very formal Navy-style dinner with the tablecloths and being served and everything. So he invites Admiral Nimitz to visit him for dinner in the next couple of days, and Admiral Nimitz shows up for his dinner, and they're sitting in a Quonset hut on a couple of crates, eating C rations, and at the end of the meal, Nimitz looks at LeMay and says, "I get your point." And then he started sending more construction materials to LeMay to help finish up the rest of the facilities.

LeMay starts by trying out his own version of his predecessor's strategy. He decides to take out the Nakajima aircraft plant in Tokyo. He needs to satisfy himself that Hansell's failure wasn't just *Hansell*.

LeMay sends his first mission against Nakajima in January, then one in February, and another in early March. Hundreds of B-29s, making the long trek to Japan. And in the end, the plant is still standing.

He has run up against the same obstacle as Hansell did. *How can I force a Japanese surrender from the air if I can't hit anything?* As Crane explains, "There's nothing else he can tweak. He says, 'Okay, I've got to try something different.'"

He starts with the wind. The jet stream is an unstoppable force. It can't be wished away, and LeMay realizes it's making everything else impossible. Precision-bombing doctrine starts with the requirement that the bomber come in high, well above the range of enemy fire and antiaircraft guns. LeMay throws that doctrine out the window. He decides the B-29s will have to come in *under* the jet stream.

Then there are the clouds. The Norden bombsight only works if the bombardier can see the target. But Japan can be almost as cloudy as England. In February of 1945, the staff meteorologists on Guam tell LeMay that he can expect no more than seven days in March when there would be skies clear enough for visual bombing. He could expect six days in April and May and four in June. How do you mount a sustained attack on Japan if you can only bomb six or seven days a month?

There's a strange stream-of-consciousness section in LeMay's autobiography where he writes:

How many times have we just died on the vine, right here on these islands? We assembled the airplanes, assembled the bombs, the gasoline, the supplies, the people. We got the crew set— everything ready, to go out and run the mission. Then what would we do? Sit on our butts and wait for the weather...So what am I trying to do now? Trying to get us to be *independent of weather.* And when we'll get ready, we'll go.

So what does "trying to get us to be independent of weather" mean? It means not only is he going to come in under the jet stream, he's also going to come in under the clouds. He's going to have the pilots come in between five thousand and nine thousand feet, lower than anyone has ever dreamed of taking a B-29 on a bombing run.

Crane explains, "Once he realizes he's going to have to go to lower altitude, then that leads to a whole set of other conclusions."

The next logical step: precision bombing was supposed to be daylight bombing. You needed to see the target before you could line up the bombsight. But if LeMay's bombers come in low during the day, they will be sitting ducks for the Japanese air defense, so he decides: *We have to come under cover of night.*

Jet stream plus heavy cloud cover means low. Low means night. And the decision to switch to night raids means you can't do precision bombing anymore—no more fiddling with the Norden, no more tight-formation flying in order to coordinate bomb strikes, no more agonizing over exactly where the target is.

And what weapon will he use for these attacks? Napalm. Napalm will work perfectly.

LeMay's anger over Schweinfurt and his frustration over the impossible conditions in India have come to a head. And so he says, there in his Quonset hut in Guam, *I'm going to do it* my *way now.* He writes out a plan for his first big attack, and instead of naming the exact target—as the Bomber Mafia would always insist on doing—he just writes: "Tokyo." Then, when he sends his plan to Washington for the approval of his boss, General Hap Arnold, he makes sure it arrives on a day when Arnold isn't in his office, "so he can get that initial raid off before Arnold really has a chance to look at it very much," Crane says. "Because he realizes he's taken a risk. B-29s are very valuable…You're talking about going in at night, low altitude. He leaves most of the ammunition and gunners behind."

The only thing LeMay lets his pilots have to defend themselves is a tail gunner. All other guns are removed. He wants to cut all excess weight so he can carry as much napalm as possible.

The airmen who flew that mission never forgot when they were first given those instructions. The B-29 airman David Braden described the briefing:

And there was just a gasp in the audience, 'cause you never thought about doing anything except high-altitude flying.

And you went out, and the bottom of your aircraft had been painted black. So you knew that this was going to be a different thing...Most of the guys thought it was a suicide mission. Some of them went in and wrote goodbye letters to their families, you know, because of the low-altitude [flight profile].

To be clear, five thousand feet is not just low. Five thousand feet is also unheard of. Twenty years later, Haywood Hansell was still astonished at the insanity of LeMay's idea:

I have been asked whether I would have done that. I think in all honesty the answer would be no. I think I'd have gone in [at] about fifteen thousand feet.

But to go in first as low as five or ten thousand feet, without any real knowledge of the density of the antiaircraft defenses, was I think a very dangerous and a very courageous thing to do if it turned out to be right, and I think that was General LeMay's personal decision.

A very dangerous and a very courageous thing. It really isn't necessary to read between the lines of what

Hansell said. The day when LeMay briefs his pilots, he almost has a mutiny on his hands. But had you confronted him that morning, he would have said, *What choice do I have?* As he put it later, "Well, I woke up one day, and I had been up there for about two months and I hadn't done anything much yet. I'd better do something."

Was he really just going to sit there and wait for the clouds to clear, the jet stream to move away, and his bombardiers to become Norden virtuosos? In an oral history recorded long after the war, he still had Haywood Hansell's disgraced exit on his mind. Here is how he responded to questions about his strategy:

Question: General LeMay, where did the idea for the low-level fire attacks originate?

LeMay: We had ideas flying back and forth, a lot. It was my basic decision. I made it…Nobody said anything about night incendiary bombing. But [we] had to have results, and I had to produce them. If I didn't produce them, or made a wrong guess, get another commander in there. That's what happened to Hansell. He got no results. You had to have them.

2.

Almost all stories in the Curtis LeMay legend are about his cold-bloodedness, his ruthlessness, his unshakable calm.

In chapter 4 of this book, I quoted him from early on in the war, after returning from a bombing mission over Europe:

Question: Colonel LeMay, how'd the trip go today?
LeMay: Well, it went pretty well, except it was rather dull compared to some that we've had. There weren't any fighters out, and flak was just moderate and very inaccurate.

He had just landed after hours of flying over enemy territory, being shot at from below and attacked from all sides by German fighter planes. *It was rather dull compared to some that we've had.*

In Europe, LeMay had insisted that his pilots not take evasive action as they flew toward their bombing targets. Every one of his pilots was terrified that if he did that, he and his crew would be gunned down by antiaircraft fire. So LeMay said, *I'll lead the first mission myself.* Remember how he later put it: "It worked out. I'll admit some uneasiness on my part and some of the other people in the outfit when we made that first straight-in bomb run, but it worked."

One of LeMay's pilots once said that when he confessed his fears to LeMay, LeMay replied: "Ralph, you're probably going to get killed, so it's best to accept it. You'll get along much better." *That's* the LeMay we know.

But every now and again, there are hints of another LeMay—for example, when he says, "I'll admit some uneasiness." That's code for *I was terrified,* but of course he couldn't let anyone see that.[*] You cannot lead airmen into battle if they can sense your fear, so terror turns into a shrug and an epic bit of understatement. LeMay was uncompromising with his men in terms of how relentlessly he prepared and drilled them, but he was that way for a *reason.* Because he cared about them. There's a line in one profile of LeMay written by St. Clair McKelway, who served under him on Guam, that I think explains this beautifully. LeMay did what he did because he had "a heart that revolted at the idea of what lack of discipline and training would mean to his young crews."

In LeMay's memoir, there's only one moment when he truly seems to let his emotional guard down. It's

---

[*] Even in letters home to his wife, LeMay was remarkably unemotional. On March 12, two days after the attack on Tokyo, he mentioned the mission only in passing: "We had a good mission to Tokyo the other day. I sent a message home to have you notified about the Army Hour program. I hope it gets there in time. I'm glad you liked the evening bag. I'm sure I spoil you. I can remember the time when that would have paid the grocery bill for a month."

when he describes the first time he saw an airplane. He
was a child, standing in the backyard of the house in
the struggling neighborhood where his family lived, in
Columbus, Ohio.

> Suddenly, in the air above me, appeared a flying-
> machine. It came from nowhere. There it was, and
> I wanted to catch it...
>
> Children can muster enormous strength in
> ideal and idea, in all their effort to grasp the
> trophy they desire. And nobody was holding me
> back, no one was standing close to say, "Look,
> you're just a little child. That airplane is away up
> there in the air, and no matter how fast you run
> you can't keep up with it. You can't reach high
> enough to seize it." I just thought that I might be
> able to grab the airplane and have it for my own,
> and possess it always. So I lit out after it.

He ran across neighbors' backyards, vacant lots,
down sidewalks. But of course he couldn't catch it.
"Then it was gone. Its wonderful sound and force and
the freakish illusion of the Thing, a Thing made of
wood and metal, piercing the air."

He went back home. And he wept.

The only time LeMay could admit to real emotion
was while telling a story from his childhood, when
the object of his affection was a mechanical device.

It is easy to understand the moral vision of someone like Haywood Hansell, or the other members of the Bomber Mafia, because they spoke the grand language of morality. Can we wage war in a way that satisfies our consciences? But LeMay is someone you have to work a little harder to understand.

LeMay's daughter, Jane LeMay Lodge, spoke about this in a 1998 oral history.

> There were a couple of very bad articles saying that he wanted to start World War III and that he was a warmonger and a hawk...Then you read an interview during the war when they did that low-level bombing—and he wasn't able to be on that mission—when he stood on that runway, counting those planes, knowing how many planes took off.
>
> Counting those planes. Standing there until the last plane is back. Now, a man who doesn't have any sensibilities and is sadistic and doesn't care where he is going or who he steps over isn't going to do that kind of thing.

So how would LeMay have justified the firebombing he intended to inflict on Japan? Well, he would have said that it was the responsibility of a military leader to make wars as short as possible. That it was the *duration* of war, not the techniques of war, that caused suffering. If you cared about the lives of your men—and the pain inflicted on your enemy—then you ought to wage as relentless and decisive and devastating a war as you could. Because if being relentless, decisive, and devastating turned a two-year war into a one-year war, wasn't that the most desirable outcome?

Satan tempts Jesus by offering him dominion over all he sees—the chance to defeat the Roman enemy—if only Jesus will accept, as one theologian puts it, "the temptation to do evil that good may come; to justify the illegitimacy of the means by the greatness of the end." Haywood Hansell sided with Jesus on that question: you should never do evil so that good may come. But LeMay would have thought long and hard about going with Satan. He would have accepted the illegitimate means if they led to what he considered a swift and more advantageous end.

As he put it years later, "War is a mean, nasty business, and you're going to kill a lot of people. No way of getting around it. I think that any moral commander tries to minimize this to the extent possible, and to me the best way of minimizing it is getting the war over as quick as possible."

That's what he said to his crews when he laid out their new mission: *What I am proposing sounds crazy, I know. But it is our only chance to end this war. Otherwise, what are our options? You want to go back to the days of Haywood Hansell, sitting on the runway, waiting for the weather to clear? We'll all be here for years then.* In Germany, the Nazis were close to surrender. The people back home in America, who had been sacrificing for four years to support the war, were exhausted. Curtis LeMay didn't think he had any time to waste. He had to act.

3.

So: Operation Meetinghouse. The night of March 9, 1945. Curtis LeMay's first full-scale attack on the city of Tokyo.

That afternoon, there was the obligatory press conference. General Lauris Norstad, the man who had sent Haywood Hansell packing, had flown in again from Washington. He and LeMay briefed the war correspondents and told them what they could and couldn't reveal. Then the planes began taking off, one by one, from the airfields on Guam, Tinian, and Saipan—more than three hundred B-29s in all, an armada. They were loaded with as much napalm as they could carry. LeMay stood on the tarmac, counting the planes.

The first bombers would not reach Tokyo until early the following morning. So for the balance of the day, there was nothing to do but wait. In the evening, LeMay went to the operations room, sat on a bench, and smoked a cigar.

St. Clair McKelway, the public relations officer on the base, found him there alone, at two in the morning. LeMay had sent everyone else home. "I'm sweating this one out myself," LeMay told McKelway. "A lot could go wrong...I can't sleep...I usually can, but not tonight."

McKelway would later write a long series for *The New Yorker* about his time with LeMay on Guam.[*] His account of that endless night of waiting is worth quoting at length:

> In deciding to send his B-29s in over Tokyo at five to six thousand feet, LeMay was increasing the risk his crews would run, and he has a deep feeling of personal responsibility for his crews; he was risking the success of the whole B-29 program, which...is dear to him in an emotional as well as an operational way; and he was risking his own future, not only, I think, as an Army officer but as a human being. If he lost seventy percent of his airplanes by such a decision, or even fifty percent of them, or even twenty-five percent of them, he

---

[*] Having left his position at *The New Yorker*, McKelway served as a lieutenant colonel in the Army. As a public relations officer, his role included censoring reports that would be damaging to his military colleagues and superiors. His postwar reporting has been sharply criticized, including at *The New Yorker* itself, for unreliable narratives and whitewashing war crimes.

would be through, and I imagine that a man like him would be through in every sense of the word, for he would have lost confidence in himself.

McKelway sat down next to LeMay on the bench. "If this raid works the way I think it will, we can shorten this war," LeMay said to McKelway. The same thing he always said. He looked at his watch. The first reports from Japan were still half an hour away.

"Would you like a Coca-Cola?" LeMay said. "I can sneak in my quarters without waking up the other guys and get two Coca-Colas and we can drink them in my car. That'll kill most of the half hour"…We sat in the dark, facing the jungle that surrounds the headquarters and grows thickest between the edge of our clearing and the sea.

The two men waited through what would turn out to be the longest night of the war.

4.

Curtis LeMay's fleet of B-29s had, as its destination, a twelve-square-mile rectangular region of central Tokyo straddling the Sumida River. It included an industrial area, a commercial area, and thousands of largely working-class homes, comprising what was, at the time, one of the most densely populated urban districts in the world.*

The first Superfortress reached Tokyo just after midnight, dropping flares to mark the target area. Then came the onslaught. Hundreds of planes—massive winged mechanical beasts roaring over Tokyo, flying so low that the entire city pulsed with the booming of their engines. The US military's worries about the city's air defenses proved groundless: the Japanese were completely unprepared for an attacking force coming in at five thousand feet.

The bombs fell from the B-29s in clusters. They were small steel pipes twenty inches long, weighing six pounds each, packed with napalm. Little baby bombs, each with a long gauze streamer at one end, so that if you looked to the sky that night in Tokyo, there would have been a moment of extraordinary beauty—thousands of these bright green daggers falling down to earth.

And then: *boom*. On impact, thousands of small explosions. The overpowering smell of gasoline. Burning globs of napalm exploding in every direction. Then

---

* As environmental historian David Fedman points out, military maps of the Tokyo attack reveal that crowded working-class civilian areas were intentionally targeted. Why? The homes of the poor were easy to light on fire: "That the more densely populated regions of the city correspond to the incendiary zone is no accident: war planners sought to exploit the vulnerability of this section of the city, composed as it was of flammable 'paper and ply-board' structures."

another wave of bombers. And another. The full attack lasted almost three hours; 1,665 tons of napalm were dropped. LeMay's planners had worked out in advance that this many firebombs, dropped in such tight proximity, would create a firestorm—a conflagration of such intensity that it would create and sustain its own wind system. They were correct. Everything burned for sixteen square miles.

Buildings burst into flame before the fire ever reached them. Mothers ran from the fire with their babies strapped to their backs only to discover—when they stopped to rest—that their babies were on fire. People jumped into the canals off the Sumida River, only to drown when the tide came in or when hundreds of others jumped on top of them. People tried to hang on to steel bridges until the metal grew too hot to the touch, and then they fell to their deaths.

Circling high above Tokyo that night was the master bomber—LeMay's deputy, Tommy Power—choreographing the attack. Historian Conrad Crane says that Power sat in his cockpit drawing pictures of everything he saw:

> [Power] remarked, "The air was so full of incendiaries you could not have walked through them."
> By 2:37, the largest visible fire area was about forty blocks long and fifteen wide. The smoke was up to twenty-five thousand feet…

When he draws his last sketch, which is about an hour after [...] his first one, there's basically a score of separate areas from fifty to a thousand city blocks burning at the same time. And his last report says that the glow from the fires was visible 150 miles away.

After the war, the United States Strategic Bombing Survey concluded the following: "Probably more persons lost their lives by fire at Tokyo in a six-hour period than at any time in the history of man." As many as one hundred thousand people died that night. The aircrews who flew that mission came back shaken.

As airman David Braden recalled, "Frankly, when those cities were on fire, it looked like you were looking into the mouth of hell. I mean, you cannot imagine a fire that big."

Conrad Crane added, "They're about five thousand feet, they are pretty low... They are low enough that the smell of burning flesh permeates the aircraft... They actually have to fumigate the aircraft when they land back in the Marianas, because the smell of burning flesh remains within the aircraft."

The next night, back on Guam, LeMay was awakened around midnight. The aerial photos taken during the attack were ready. As news spread, people came running from their beds. They drove up in Jeeps until the room was crowded. LeMay, still in his pajamas, put the photos down on a large table under a bright light. There was a moment of shocked silence. St. Clair McKelway was standing in the room with all the others and remembers LeMay gesturing at the vast area of devastation. "All this is out." LeMay said. "This is out—this—this—this."

General Lauris Norstad stood next to him and said, "It's all ashes—all that and that and that."[*]

---

[*] Despite the incalculable loss of life, there remains no government-sanctioned memorial in Japan to the March 9 attack. Survivors of that night, who call themselves "memory activists," have struggled to commemorate the Tokyo raid in the face of political and public apathy. Eventually they funded their own memorial—the Center of the Tokyo Raids and War Damage. In his forthcoming documentary, *Paper City*, director Adrian Francis interviews survivors of the 1945 firebombing of Tokyo to preserve their stories and their fight for remembrance.

CHAPTER NINE

"*Improvised destruction.*"

1.

After the firebombing of Tokyo in March of 1945, Curtis LeMay and the Twenty-First Bomber Command ran over the rest of Japan like wild animals. Osaka. Kure. Kobe. Nishinomiya. LeMay burned down 68.9 percent of Okayama, 85 percent of Tokushima, 99 percent of Toyama—sixty-seven Japanese cities in all over the course of half a year. In the chaos of war, it is impossible to say how many Japanese were killed— maybe half a million. Maybe a million. On August 6, the *Enola Gay,* a specially outfitted B-29, flew from the Marianas to Hiroshima and dropped the world's first atomic bomb. Yet LeMay kept going. In his memoirs, the nuclear attacks get no more than a couple of pages. That was someone else's gig.

> Our B-29s went to Yawata on August 8th and burned up 21 percent of the town, and on the same day some other B-29s went to Fukuyama and burned up 73.3 percent. Still there wasn't any gasp and collapse when the second nuclear bomb went down above Nagasaki on August 9th. We kept on flying. Went to Kumagaya on August 14th…45 percent of that town. Flew our final mission the same day against [Isesaki], where we burned up 17 percent of that target. Then the crews came home to the Marianas and were told that Japan had capitulated.

LeMay always said that the atomic bombs were superfluous. The real work had already been done.

2.

There is a story that LeMay loved to tell about his firebombing campaign. It's in his memoirs and in interviews he gave after his retirement. And each time he told the story, the language—the phrases, the order of details—is the same, as if it were part of his repertoire. It involved a fellow general named Joseph Stilwell.

Stilwell was the head of US operations in the China-Burma-India theater. He was a generation older than LeMay. He was traditional Army, out of West Point. His nickname was Vinegar Joe. He was shrewd and ornery. On his desk was a plaque with a mock-Latin inscription—*Illegitimi non carborundum.* "Don't let the bastards grind you down." Of course LeMay wanted to meet Stilwell, so one day he paid him a courtesy call.

As LeMay told the story:

> I went up to New Delhi to call on him. He was out in the jungle someplace. Well, I wasn't about to go run him down in the jungle. I just left a card, and saw the chief of staff, and went home.

A very LeMay beginning to the story: a little belligerence. *I wasn't about to go run him down in the jungle.* LeMay tried again, and not long thereafter he met up with Stilwell at the B-29 staging base in China, in Chengdu. LeMay wanted to show Stilwell what the Twentieth Bomber Command was up to.

> I took him in tow with me, and we got the mission off, and then had dinner, and [I] stayed up all night talking to him, trying to explain to him what strategic bombardment was all about, and what we were trying to do, and how we were going about doing it, and so forth…I couldn't

get to first base. Just couldn't, literally couldn't get to first base.

In other words, he couldn't make himself clear.

There they are, two distinguished generals, having dinner and drinks in the middle of China. And LeMay is trying to explain to his colleague what he's doing, what he wants to do, what he thinks can be accomplished with this marvelous new plane called the B-29. He was trying to communicate the idea that airpower did not have to be used specifically in support of ground troops—that you had other options. That airpower could leapfrog over the front lines of battle and attack behind the lines. It could take out manufacturing plants, power grids, and entire cities if you wanted.

Did he talk about napalm? He must have. The work on the replica Japanese buildings in the Utah desert was a matter of record. And LeMay had already used napalm at least once, on one of his bombing runs into Japan. So maybe he went even further and said to Stilwell, *You know, we could just burn the whole country down.*

And Stilwell—as savvy and experienced and grizzled a military mind as there was in the Second World War—hadn't the slightest clue what LeMay was talking about. What did this mean? You would wage an entire war from the sky?

A year passes. Japan surrenders, and the two men meet up again.

And the next time I saw him was when we went out to the *Missouri* in Yokohama. For the surrender, he was there. And when we went into Yokohama— Yokohama was a city of about four and a half million then, I guess—I didn't see a hundred Japs in Yokohama. I'm sure there were more than that around, but they stayed out of sight.

LeMay had hit Yokohama in May of 1945, two months after Tokyo. More than 450 B-29s dropped 2,570 tons of napalm, reducing half the city to ashes and killing tens of thousands. A couple of days after their surrender-day encounter in Yokohama, LeMay and Stilwell met again in Guam. As LeMay later recalled:

[Stilwell] came over to see me, and he said, "LeMay, I stopped to tell you that it finally dawned on me what you were talking about...And it didn't dawn on me until I saw Yokohama."

Why didn't Stilwell understand, back in that first conversation in China, what LeMay was intending? It's not like Stilwell was some shrinking violet. When he walked around the rubble of Yokohama, he was delighted. This is what he wrote in his diary: "What a kick to stare at the arrogant, ugly, moon-faced, buck-toothed, bowlegged bastards, and realize where this puts them. Many newly demobilized soldiers around.

Most police salute. People generally just apathetic. We gloated over the destruction & came in at 3:00 feeling fine."

That's the kind of man Stilwell was. Yet he had to see, with his own eyes, what airpower *did* to Yokohama to understand LeMay, because what LeMay had been talking about in their conversation in China was outside the old general's imagination. He had been taught back at West Point that soldiers fought soldiers and armies fought armies. A warrior of Stilwell's generation was slow to understand that you could *do* this, as an American Army officer, if you wanted: you could take out entire cities. And then more. One after another.

Roosevelt's secretary of war, Henry Stimson, reacted the same way. Stimson was responsible, more than anyone, for the extraordinary war machine that the United States built in the early years of the Second World War. He was a legend, the eldest of the elder statesmen, a blue blood, the adult in the room during any discussion of military strategy or tactics. But he seemed strangely oblivious to what his own air forces were up to.

General Hap Arnold, head of the Army Air Forces, once told Stimson, with a straight face, that LeMay was trying to keep Japanese civilian casualties to a minimum. And Stimson believed him. It wasn't until LeMay firebombed Tokyo a second time, at the end of May, that Stimson declared himself shocked at what was happening in Japan. Shocked? This was two and a half months after LeMay had incinerated sixteen square miles of Tokyo the first time around.

Historians have always struggled to make sense of Stimson's obliviousness.* Military historian Ronald Schaffer writes in his book *Wings of Judgment,*

> Was it possible that the secretary of war knew less about the March 10 bombing of Tokyo than a reader of the *New York Times*? Why did he accept Arnold's statement about attempting to limit the impact of bombing on Japanese civilians? Was he signaling that he really did not wish to be told what the AAF was doing to enemy civilians?

I wonder if the explanation for Stimson's blindness isn't the same as the explanation for Stilwell's. What LeMay was doing that summer was simply outside his imagination.

When we talk about the end of the war against Japan, we tend to talk about the atomic bombs dropped on

---

* Stimson leaves behind a complicated legacy. In private writings, he expresses concern over the potential loss of civilian life and opposes the destruction of cultural centers such as Kyoto. But as historians have noted, Stimson's delusions around the incendiary bombing campaign seem inexcusable, if not totally implausible. In the eastern front, after a particularly damaging AP report that cited American commanders' plans to conduct "deliberate terror bombing of the great German population centers as a ruthless expedient to hasten Hitler's doom," Stimson sought to spin the narrative in his favor: "Our policy never has been to inflict terror bombing on civilian populations."

Nagasaki and Hiroshima in August of 1945. The use of nuclear weapons against Japan was a matter of serious planning and consideration. It was endlessly debated and agonized over at the highest levels. *Should we use the bomb? If so, where? Once? Twice? Have we set a dangerous precedent?* President Truman, who had taken office after Roosevelt died, in the spring of 1945, was advised by a panel of military and scientific experts, weighing the decision well in advance. Truman lost sleep over the decision. He wandered the halls of the White House.[*]

But LeMay's firebombing campaign unfolded with none of that deliberation. There was no formal plan behind his summer rampage, no precise direction from his own superiors. To the extent that the war planners back in Washington conceived of a firebombing campaign, they thought of hitting six Japanese cities, not sixty-seven. By July, LeMay was bombing minor Japanese cities that had no strategically important industry at all—just people, living in tinderboxes. The historian William Ralph calls LeMay's summer bombing campaign "improvised destruction":

---

[*] In his diary on July 25, 1945, Truman wrote: "We have discovered the most terrible bomb in the history of the world…This weapon is to be used against Japan between now and August 10th. I have told the Sec. of War, Mr. Stimson, to use it so that military objectives and soldiers and sailors are the target and not women and children. Even if the Japs are savages, ruthless, merciless and fanatic, we as the leader of the world for the common welfare cannot drop that terrible bomb on the old capital or the new."

It is striking that such a lethal campaign...sprang from the commander in the field. How was it permitted to originate this way? How could a decision laden with such ethical and political consequences be handed to a young field commander? Where was the personal responsibility and active involvement from above?

But up above, people like Stimson and Stilwell could not—or would not—wrap their minds around what LeMay was doing. They struggled not just with the *scale* of the destruction LeMay planned and inflicted on Japan that summer but also with the audacity of it. A man, out there in the Marianas, falls in love with napalm, comes up with an improvised solution to get around the weather. And then he just keeps going and going.

3.

The ground invasion of Japan—which both the Japanese and American militaries dreaded—never had to happen. In August of 1945, Japan surrendered. This was exactly the outcome LeMay had hoped for that night in March, after he sent his first armada of B-29s to Tokyo. He had sat in his car with St. Clair McKelway and said, "If this raid works the way I think it will, we can shorten this war." You wage war as ferociously and brutally as possible, and in return, you get a shorter war.

The historian Conrad Crane told me:

I actually gave a presentation in Tokyo about the incendiary bombing of Tokyo to a Japanese audience, and at the end of the presentation, one of the senior Japanese historians there stood up and said, "In the end, we must thank you, Americans, for the firebombing and the atomic bombs."

That kind of took me aback. And then he explained: "We would have surrendered eventually anyway, but the impact of the massive firebombing campaign and the atomic bombs was that we surrendered in August."

In other words, this Japanese historian believed: no firebombs and no atomic bombs, and the Japanese don't surrender. And if they don't surrender, the Soviets invade, and then the Americans invade, and Japan gets carved up, just as Germany and the Korean peninsula eventually were.

Crane added,

The other thing that would have happened is that there would have been millions of Japanese who would have starved to death in the winter. Because what happens is that by surrendering in August, that gives MacArthur time to come in with his occupation forces and actually feed Japan…I mean, that's one of MacArthur's great successes: bringing in a massive amount of food to avoid starvation in the winter of 1945.

He is referring to General Douglas MacArthur, the supreme commander for the Allied powers in the Pacific. He was the one who accepted the Japanese emperor's surrender.

Curtis LeMay's approach brought everyone—Americans and Japanese—back to peace and prosperity as quickly as possible. In 1964, the Japanese government awarded LeMay the highest award their country could give a foreigner, the First-Class Order of Merit of the Grand Cordon of the Rising Sun, in appreciation for his help in rebuilding the Japanese Air Force. "Bygones are bygones," the premier of Japan said at the time, dismissing the objections of his colleagues in the Japanese parliament. "It should be but natural that we reward the general with a decoration for his great contribution to our Air Self-Defense Units."

Somewhere in retirement, Haywood Hansell saw that announcement in the newspaper, and I'm sure he wondered why he didn't get an award as well for the effort he put toward fighting a war with as few civilian casualties as possible. But we don't give prizes to people who fail at their given tasks, no matter how noble their intentions, do we? To the victor go the spoils.

But if Curtis LeMay won the war and the prizes, why is it that Haywood Hansell's memory is the one that moves us? Romantic, idealistic Haywood Hansell, who loved Don Quixote, who identified with the delusional gallant knight who tilted helplessly at windmills. We can admire Curtis LeMay, respect him, and try to understand his choices. But Hansell is the one we give our hearts to. Why? Because I think he provides us with a model of what it means to be *moral* in our modern world. We live in an era when new tools and technologies and innovations emerge every day. But the only way those new technologies serve some higher purpose is if a dedicated band of believers *insists* that they be used to that purpose. That is what the Bomber Mafia tried to do—even as their careful plans were lost in the clouds over Europe and blown sideways over the skies of Japan. They persisted, even in the face of technology's inevitable misdirection, even when abandoning their dream offered a quicker path to victory, even when Satan offered them all the world if only they would renounce their faith. Without persistence, principles are meaningless. Because one day your dream may come true. And if you cannot keep that dream alive in the interim, then who are you?

I asked the military historian Tami Biddle, who teaches at the Army War College, what she tells her students about the spring and summer of 1945, and she recounted a personal story. "My grandmother Sadie Davis had two children, two sons fighting in World War II. One had been in the Pacific theater for a long time; one had been fighting in the European theater but didn't have enough points to leave the war prior to what would have been the landing on Kyushu."

The landing on Kyushu was the planned invasion of Japan in November of 1945, an invasion expected to cost the lives of more than half a million American soldiers, not to mention just as many Japanese. She continued,

> He would have been in that landing had it not been for the Americans being exceedingly brutal with the Navy and the blockade, with the air war against Japanese cities, and then, ultimately, with atomic weapons.*
>
> For her, I'm sure that she was quite prepared for us to be brutal in that moment, because she wanted her sons to come home. Lots of people feel that way in wartime. After the war, you look

---

* George C. Marshall, the General of the Army, believed that dragging out the war would destroy morale. He argued that the fastest path to victory was an amphibian land invasion of Japan. By contrast, Fleet Admiral Ernest J. King, who led the Navy, believed that a land invasion risked far too many casualties. Ultimately, these plans were never fully realized. Japan surrendered before the Navy blockade was expanded, and the land invasion, dubbed Operation Downfall, never commenced.

at the whole situation and the totality of the thing, and you look at what has been wrought, and you look at the lives lost and the devastation and the pictures of Hiroshima and the pictures of the cities that were bombed in Germany. You think, "Dear God, was there some other way? Did we lose our souls? Did we go into a Faustian bargain to win, where winning cost us so much morally?"

Curtis LeMay put the bomb-damage photos of Schweinfurt and Regensburg in the foyer of his house because he wanted to remind himself every day of how many of his men were lost in the course of what he considered a fruitless mission. I would feel better about Curtis LeMay if he had also hung the strike photos from the firebombing of Tokyo—to remind himself, every day, of what was lost in the course of what he considered his most successful mission.*

---

* In the end, history probably best remembers Curtis LeMay for a remark he made in his memoirs, published just before his retirement, in 1965. LeMay is quoted as saying this about North Vietnam: "We're going to bomb them back into the Stone Age." This remark was featured in media coverage when LeMay ran for vice president on a third-party ticket with the segregationist George Wallace, in 1968. But a 2009 biography of LeMay by Warren Kozak calls the truth of this famous quotation into question. Kozak writes: "In his autobiography, *Mission with LeMay*, written with the help of novelist MacKinley Kantor, LeMay gave Kantor his quotes, stories, and ideas, and Kantor helped shape them into written form. The drafts of the book were sent to LeMay for his approval before it was published. The book is very much in LeMay's voice, and it is well done. But

As Biddle says,

Those are really unresolvable questions. I hope
I never have to face the circumstances that my
grandmother faced having two sons in a war and
having to maybe hope for the kinds of things
that she was hoping for—devastating attacks on
an enemy that would finally make the war end so
that her boys could come home. I hope I never
have to face that in my lifetime. I'm reluctant to
judge the people who feel that way.

---

there is one quote on page 545 concerning Vietnam that Kantor
invented: 'My solution to the problem would be to tell them frankly
that they've got to draw in their horns and stop their aggression,
or we're going to bomb them back into the Stone Age. And we
would shove them back into the Stone Age with Air power or Naval
power—not with ground forces.' To this day, when LeMay's name
comes up, most people remember that quote, asking 'Isn't he the guy
who wanted to bomb Vietnam back to the Stone Age?' Much later,
LeMay admitted to friends that he never said those words. 'I was just
[so] damned bored going through the transcripts that I just let it get
by,' he told friends and family. Since he put his name on the book, he
was responsible, but the quote most likely stayed with him simply
because it sounded like something he could have said."

"All of a sudden, the Air House would be gone. Poof."

When I was writing *The Bomber Mafia*, I spent an evening at the Air House, in Fort Myer, across the Potomac River from Washington, DC. It is the official residence of the chief of staff of the Air Force. I mentioned this night at the beginning of this book. The then Air Force chief of staff, General David Goldfein, invited me to sit and talk with a group of his fellow Air Force generals.

The Air House is on a street lined with gracious Victorian homes. The head of the Joint Chiefs of Staff lives on that street. The vice chairman of the Joint Chiefs, who joined us as well, lives next door. Across the street is the field where the Wright brothers gave their first aerial demonstration to the Army brass. Inside the house, on a wall of the dining room, are photographs, arranged in order, of everyone who has occupied the top post at the Air Force since it was established as a separate service, in 1947. I stood in front of those photographs for a long time, looking at all the names and faces I had been reading and hearing about. And in the top row, fifth from the left, was Curtis LeMay, scowling at the camera.*

It was a hot summer night. We sat outside in deck chairs — five of us. Planes roared overhead as they took off from nearby Reagan National Airport. A big air-conditioning unit hemmed and hawed. Mosquitoes buzzed about happily. And the generals talked about the wars they fought: Kosovo. Desert Storm. Afghanistan. Some of them had fathers who served in Vietnam and grandfathers who served in World War II — so they had a sense, a personal sense, of how things used to be and how things have changed.

One of the generals told a story about his time in western Afghanistan. He'd gotten a call from a group of soldiers. They'd been attacked.

I've got a guy on the ground talking to me on a radio, and you can hear the fifty-caliber machine guns going off all around him. He says, "I'm

---

* LeMay took over as head of the Strategic Air Command in 1948. As historian Richard Kohn notes, "General LeMay, more than any other figure, shaped the Strategic Air Command (SAC) during its formative years under his command (1948–57)." In 1961, LeMay rose even higher when President Kennedy made him Air Force chief of staff.

surrounded on three sides. I'm taking effective fire. I've got guys wounded in my compound. We're going to get overrun."

The troops on the ground needed air cover. But if the bomb missed by even ten yards, it would take out the US troops. He goes on: "So three different bombs [land] within twenty meters of this guy, taking out three different buildings, and the guy [survives] with his team. That's how precise precision-guided bombs can be."

Goldfein pointed to the long rows of homes on either side of the Air House. He said his father, who flew an F-4 fighter jet in Vietnam, could have dropped six bombs on that street and been reasonably sure that at least one or two would hit the Air House. By contrast, Goldfein said, "his son rolls into Desert Storm, and I can tell you...that with 89 percent confidence I'm going to have my bombs hit that building."

But just a few years after the US invasion of Kuwait, General Goldfein was leading a squadron into Kosovo. And by that point, he said, he would have been confident that he could take out not simply the Air House but a specific wing of the Air House.

Okay, so now you roll forward from then to today. Today, the expectation is that a young pilot can hit just above the pinnacle at the base of the chimney. And...if he didn't hit that, then that's a miss. That level of precision. And...the reason I use that as an example is that the target is an individual who's in that room. And I don't want to destroy the floors below it. We do that all the time. That's the level of precision we've achieved.

None of the generals that evening claimed that this precision-bombing revolution had perfected war, or solved war. It has its own set of drawbacks. If your target is a single man inside a room, then you have to have intelligence good enough to tell you that this is the man you want. And when you have a way of hitting a man inside a room, then it becomes awfully easy to decide to strike, doesn't it? They all worried about that fact: the cleaner and more precise a bomber gets, the more tempting it is to use that bomber—even when you shouldn't.

Still, think about this. In 1945, someone who wanted to take out that house Goldfein was pointing at might have come in with an armada of bombers, a few thousand tons of napalm, and burned everything to the ground for miles around—Washington, DC, across the river; Arlington, Virginia, on the other side of the base.

There is a set of moral problems that can be resolved only with the application of conscience and will. Those are the hardest kinds of problems. But there are other problems that can be resolved with the application of human ingenuity. The genius of the Bomber Mafia was to understand that distinction—and to say, *We don't have to slaughter the innocent, burn them beyond recognition, in pursuit of our military goals. We can do better.* And they were right.[*]

The generals began to talk about the B-2 bomber— the Stealth Bomber—the modern-day Air Force's equivalent of Curtis LeMay's B-29. But this time, with the power to come out of nowhere, undetectable.

One general said, "So in essence, [in] Fort Myer, where we're sitting today: you could take the eighty targets you want, and so from above forty thousand feet without seeing it, without [the bomber's] being on your radar, those just go away." I asked whether we would be able to hear the bomber's approach. The reply: "You don't. It's too high. You don't hear it."

We would all be sitting in our deck chairs in the backyard, and we would look up, and all of a sudden, the Air House—or maybe even some specific part of the Air House—would be gone. Poof.

High-altitude precision bombing.

Curtis LeMay won the battle. Haywood Hansell won the war.

---

[*] On January 21, 2009, the day after his inauguration, President Obama signed a United Nations protocol banning the use of incendiary weapons. As of this writing, 115 nations have signed the disarmament treaty, first introduced in 1981.

Acknowledgments

The Bomber Mafia had an unusual birth, because it began its literary life as an audiobook and then was transformed into print. Most books have the opposite trajectory. So my thanks begin with the team at Pushkin Industries who helped me create this book in its original form: Brendan Francis Newnam and Jasmine Faustino, who oversee Pushkin Audiobooks; my editor, Julia Barton; my producers, Jacob Smith and Eloise Lynton; my fact-checker, Amy Gaines; composer Luis Guerra; and sound and engineering wizards Flawn Williams and Martín H. Gonzalez. Thanks also to researchers past and present, including Camille Baptista, Stephanie Daniel, Beth Johnson, and Xiomara Martinez-White— not to mention Heather Fain, Carly Migliori, and Mia Lobel.

The crew at Little, Brown, who has been my publisher since the very beginning of my book-writing career, then took over from the audiobook team at Pushkin. I'm grateful to the following people at Little, Brown for their help in making *The Bomber Mafia* into a printed book and ebook: Bruce Nichols, Terry Adams, Massey Barner, Pam Brown, Judy Clain, Barbara Clark, Sean Ford, Elizabeth Garriga, Evan Hansen-Bundy, Pat Jalbert-Levine, Gregg Kulick, Miya Kumangai, Laura Mamelok, Asya Muchnick, Mario Pulice, Mary Tondorf-Dick, and Craig Young.

And last but not least, both General David L. Goldfein and General Charles Q. Brown Jr., the twenty-first and twenty-second chiefs of staff of the Air Force, respectively, were immensely generous in giving me guidance and access to the Air Force archives and the historians at Air University.

While I was in the middle of writing this book, General Goldfein retired and was replaced by General Brown. I watched the ceremony online. Everyone from the secretary of defense to the chairman of the Joint Chiefs of Staff—all the way on down—spoke. In the middle of one of the most tumultuous and uncertain summers in recent American history, the changeover ceremony was a model of grace, decorum, and gravitas. The original Bomber Mafia helped build one of the truly great American institutions. Their influence has not ended.

Notes

Quotations from the following sources were taken from author interviews:

Tami Biddle	John M. Lewis
Conrad Crane	Stephen L. McFarland
David Goldfein	Richard Muller
Robert Hershberg	Robert Neer
Ken Israel	Robert Pape
Richard Kohn	

INTRODUCTION: "THIS ISN'T WORKING. YOU'RE OUT."

"With 2,200 ... have to do" and "B-29s on Saipan ... its first target": William Keighley, dir., *Target Tokyo* (Culver City, CA: Army Air Forces First Motion Picture Unit, 1945), available at https://www.pbs.org/wgbh/americanexperience/features/pacific-target-tokyo/.

"I wonder if...period of years": Sir Arthur Harris, *Bomber Offensive* (London: Collins, 1947; Barnsley, UK: Pen & Sword, 2005), 72–73. Citations refer to the Pen & Sword edition.

"I thought the earth...crushed" and "Old pilots...away-y-y-y": Charles Griffith, *The Quest: Haywood Hansell and American Strategic Bombing in World War II* (Montgomery, AL: Air University Press, 1999), 189, 196.

"General...picture is taken" and "Where...stand": St. Clair McKelway, "A Reporter with the B-29s: III—The Cigar, the Three Wings, and the Low-Level Attacks," *The New Yorker*, June 23, 1945, 36.

CHAPTER ONE: "MR. NORDEN WAS CONTENT TO PASS HIS TIME IN THE SHOP."

"Mr. Norden...eighteen-hour day": Albert L. Pardini, *The Legendary Norden Bombsight* (Atglen, PA: Schiffer Publishing, 1999), 51.

"read Dickens avidly...the simple life": Stephen L. McFarland, *America's Pursuit of Precision Bombing, 1910–1945* (Washington, DC: Smithsonian Institution Press, 1995), 52.

"In the hands...a killer": Robert Jackson, *Britain's Greatest Aircraft* (Barnsley, UK: Pen & Sword, 2007), 2.

"One fellow...getting into": Donald Wilson, interview by Hugh Ahmann for the United States Air Force Oral History Program, Carmel, CA, December 1975, Donald Wilson Papers, George C. Marshall Foundation, Lexington, VA.

"Then out of nowhere...'I had a dream'" and "I had a dream...sue for peace": Donald Wilson, *Wooing Peponi: My Odyssey Through Many Years* (Monterey, CA: Angel Press, 1973), 237.

"One of them...your Norden bombsight" and "Now look at...same direction": *Principles of Operation of the Norden Bombsight,* US Army Air Forces training movie 23251, available at https://www.youtube.com/watch?app=desktop&feature=share&v=143vi97a4tY.

"I solemnly swear...life itself": *Bombs Away,* yearbook of the bombardier training school, class of 1944–46, Victorville Army Air Field, Victorville, CA, 16, available at http://www.militarymuseum.org/Victorville%20AAF%2044-6.pdf.

CHAPTER TWO: "WE MAKE PROGRESS UNHINDERED BY CUSTOM."

For information about the feminist movement in the 1970s, see Jill Lepore, *These Truths: A History of the United States* (New York: W. W. Norton, 2018), 652.

"can of its own...the future" and "If success...other combat arms": General John J. Pershing to General Charles T. Menoher, January 12, 1920, quoted in *Report of the Director of Air Service to the Secretary of War* (Washington, DC: Government Printing Office, 1920), 11.

"We were highly...rest of the Navy" and "Nobody seemed...we were giving": Harold George, interview for the United States Air Force Oral History Program, October 23, 1970, Clark Special Collections Branch, McDermott Library, United States Air Force Academy, Colorado Springs, CO.

"I feel quite certain...us to do it" and **"Now, when we began...industrial area":** Donald Wilson, interview by Hugh Ahmann for the United States Air Force Oral History Program, December 1975, Donald Wilson Papers, George C. Marshall Foundation, Lexington, VA.

"This is a quiet place...them up": Carl H. Builder, *The Masks of War: American Military Styles in Strategy and Analysis* (Baltimore, MD: Johns Hopkins University Press, 1989), 34.

"I came home...tetrahedrons together" and **"You get into technology...major problem":** Edited excerpt from longer interview conducted by Betty J. Blum in 1995 for the Chicago Architects Oral History Project, organized by the Department of Architecture at the Art Institute of Chicago. Oral history of Walter Netsch/interviewed by Betty J. Blum, compiled under the auspices of the Chicago Architects Oral History Project, the Ernest R. Graham Study Center for Architectural Drawings, Department of Architecture, the Art Institute of Chicago. Copyright 1997–2000, the Art Institute of Chicago; used with permission.

"We see then...outside power!": Phil Haun, ed., *Lectures of the Air Corps Tactical School and American Strategic Bombing in World War II* (Lexington, KY: University Press of Kentucky, 2019), Google Books.

CHAPTER THREE: "HE WAS LACKING IN THE
BOND OF HUMAN SYMPATHY."

Ira Eaker's quotations in this chapter, unless otherwise
indicated, are from interviews with Generals Ira Eaker,
Curtis LeMay, James Hodges, James Doolittle, Barney
Giles, and Edward Timberlake, recorded 1964, Air
Force Historical Research Agency, Montgomery, AL, at
http://airforcehistoryindex.org/data/001/019/301.xml.

"London raises...every morning": Humphrey Jennings and
Harry Watt, dirs., *London Can Take It!* (London: GPO
Film Unit, Ministry of Information, 1940), available at
https://www.youtube.com/watch?v=bLgfSDtHFt8.

"We used to go...the glass at the same time": Elsie Elizabeth
Foreman oral history, December 1999, Imperial War
Museums, London, available at https://www.iwm.org.uk/
collections/item/object/80018439. Reference number 19924.

"No. I never thought...will be England again": Sylvia Joan
Clark oral history, June 2000, Imperial War Museums,
London, available at https://www.iwm.org.uk/collections/
item/object/80019086. Reference number 20305.

"I asked Harris...from the British": James Parton, *Air Force
Spoken Here: General Ira Eaker and the Command of
the Air* (Montgomery, AL: Air University Press, 2000),
152–53.

Material about Frederick Lindemann, his friendship with
Churchill, and C. P. Snow's lectures was featured in
a 2017 episode of the *Revisionist History* podcast, "The

Prime Minister and the Prof" (http://revisionisthistory.com/episodes/15-the-prime-minister-and-the-prof). Quotations from C. P. Snow are from "Science and Government" (Godkin Lecture Series at Harvard University, November 30, 1960), GBH Archives. For information about Churchill and his spending on alcohol, see David Lough's *No More Champagne: Churchill and His Money* (New York: Picador, 2015), 240.

For more information on the concept of transactive memory, see Daniel M. Wegner, Ralph Erber, and Paula Raymond, "Transactive Memory in Close Relationships," *Journal of Personality and Social Psychology* 61, no. 6 (1991): 923–29, available at http://citeseerx.ist.psu.edu/viewdoc/download?doi=10.1.1.466.8153&rep=rep1&type=pdf.

"He would not shrink...professional opponents": Frederick Winston Furneaux Smith, Earl of Birkenhead, *The Prof in Two Worlds: The Official Life of Professor F. A. Lindemann, Viscount Cherwell* (London: Collins, 1961), 116.

"He was indeed...relationship with him" and **"I define...to my friends":** Roy Harrod, *The Prof: A Personal Memoir of Lord Cherwell* (London: Macmillan, 1959), 72, 73.

"The Nazis entered...reap the whirlwind": *Defence: World War II; Air Marshal Harris on Bombing Raids,* Reuters via British Pathé, British Paramount newsreel, 1942, available at https://youtu.be/fdoUZtCbsW8?t=32.

Information about the impact of the bombing on Cologne is from Max G. Tretheway, "1,046 Bombers but Cologne Lived," *New York Times,* June 2,

1992, available at https://www.nytimes.com/1992/06/02/opinion/IHT-1046-bombers-but-cologne-lived.html.

"Sir, you are…kill people: Germans": Henry Probert, *Bomber Harris: His Life and Times; The Biography of Marshal of the Royal Air Force Sir Arthur Harris, Wartime Chief of Bomber Command* (London: Greenhill Books, 2001), 154–55.

"Well, of course people…army advances" and **"We weren't aiming…draw the line?":** Arthur Harris, interview by Mark Andrews, British Forces Broadcasting Service, 1977, Imperial War Museums, London, available at https://www.iwm.org.uk/collections/item/object/80000925. Reference number 931.

CHAPTER FOUR: "THE TRUEST OF THE
TRUE BELIEVERS."

"What in heaven's…your son": Charles Griffith, *The Quest: Haywood Hansell and American Strategic Bombing in World War II* (Montgomery, AL: Air University Press, 1999), 34. Hansell singing "The Man on the Flying Trapeze" to his men is described on page 120, and the story about Hansell's meeting and courtship of his wife is told on pages 32–33.

"We have not put…problems to solve": Ralph H. Nutter, *With the Possum and the Eagle: The Memoir of a Navigator's War Over Germany and Japan* (Denton, TX: University of North Texas Press, 2005), 216.

"In short...reality in it": Miguel de Cervantes, *The Ingenious Gentleman Don Quixote of La Mancha, Volume 1,* trans. John Ormsby (London: Smith, Elder & Co., 1885), available at https://www.gutenberg.org/files/5921/5921-h/5921-h.htm.

"Well, the selection...capacity of Germany" and "There was a rain...ball-bearing industry": Haywood Hansell, talk at the United States Air Force Academy, April 19, 1967, Clark Special Collections Branch, McDermott Library, United States Air Force Academy, Colorado Springs, CO.

Quotations from the 1943 interview with Curtis LeMay are from *First U.S. Raid on Germany,* Reuters, British Pathé newsreel, 1943, available at https://www.youtube.com/watch?v=YgO6DX_9z0I.

"The briefings lasted...only comment": Russell E. Dougherty, interview by Alfred F. Hurley, Arlington, VA, May 24, 2004, University of North Texas Library, Denton, TX, available at https://digital.library.unt.edu/ark:/67531/metadc306813/.

The following quotations are from the Curtis LeMay oral history interview of March 1965, Air Force Historical Research Agency, Montgomery, AL, at http://airforcehistoryindex.org/data/001/000/342.xml: "One of the things...not very good"; "Not only were...over to the Continent"; "Something had to be done...bombsight level"; "All of the people...shoot you down"; "It required I think...too bad to me"; and "I'll admit some uneasiness...but it worked."

"He was the finest...kind of commander he was": Errol Morris, dir., *The Fog of War: Eleven Lessons from the Life of Robert S. McNamara* (New York: Sony Pictures Classics, 2003).

The following quotations are from Reminiscences of Curtis E. LeMay: Oral History, 1971 (Air Force Academy Project, Columbia Center for Oral History, Columbia University Libraries, New York, NY): "The Air Force has been battling...Find the battleship"; "Finally they agreed...this time, I'm sure"; and "Everybody [was] diving...hurt a little bit."

"I remember watching...frag like that": Curtis E. LeMay with MacKinlay Kantor, *Mission with LeMay: My Story* (New York: Doubleday, 1965), 150.

"Dawn, August seventeenth...deep in Germany" and "By the time we...dispatched to date": *The Air Force Story: Chapter XIV — Schweinfurt and Regensburg, August 1943*, produced by the Department of the Air Force, 1953, available at https://www.youtube.com/watch?v=dB8C -CagZeU.

CHAPTER FIVE: "GENERAL HANSELL WAS AGHAST."

For more information about the Schweinfurt-Regensburg raid, see Thomas M. Coffey, *Decision Over Schweinfurt: The U.S. 8th Air Force Battle for Daylight Bombing* (New York: David McKay, 1977).

Curtis LeMay's quotations in this chapter, unless otherwise noted, are from Reminiscences of Curtis E. LeMay: Oral History, 1971 (Air Force Academy Project, Columbia Center for Oral History, Columbia University Libraries, New York, NY).

"A shining silver…100 percent loss": Lieutenant Colonel Beirne Lay Jr., "I Saw Regensburg Destroyed," *Saturday Evening Post,* November 6, 1943.

"Göring's Luftwaffe…explosives to deliver" and "After getting eighty…home fast": *The Air Force Story: Chapter XIV—Schweinfurt and Regensburg, August 1943,* produced by the Department of the Air Force, 1953, available at https://www.youtube.com/watch?v=dB8C-CagZeU.

The information about the condition of the Kugelfischer ball-bearing plant after the air attack is from Thomas M. Coffey, *Decision Over Schweinfurt: The U.S. 8th Air Force Battle for Daylight Bombing* (New York: David McKay, 1977), 81.

"there is no evidence…war production": *The United States Strategic Bombing Survey: Summary Report: European War,* September 30, 1945, 6, available at https://www.google.com/books/edition/The_United_States_Strategic_Bombing_Surv/EfEdkyz_D0AC?hl=en&gbpv=1.

"There's only one…fly 'em some more": Henry King, dir., *Twelve O'Clock High* (Los Angeles: 20th Century Fox, 1949).

"We were assigned…for that squadron": National WWII Museum, *George Roberts 306th Bomb Group,* available at https://www.youtube.com/watch?v=fRO1R7Op1ec.

"We took off...aiming point": Alan Harris, ed., "The 1943 Munster Bombing Raid in the Words of B-17 Pilot Keith E. Harris (1919–1980)," AlHarris.com, available at http://www.alharris.com/stories/munster.htm.

The anecdote about the navigator who faced court-martial is from Seth Paridon, "Mission to Munster," National WWII Museum, November 20, 2017, available at https://www.nationalww2museum.org/war/articles/mission-munster; and Ian Hawkins, *Munster: The Way It Was* (Robinson Typographics, 1984), 90.

"General Hansell was aghast": Ralph H. Nutter, *With the Possum and the Eagle: The Memoir of a Navigator's War Over Germany and Japan* (Denton, TX: University of North Texas Press, 2005), 137.

"The idea...pervasive thing," "We were reasonably... personal salvation," and **"One of the things...difficult to do":** Leon Festinger, interview by Dr. Christopher Evans for the Brain Science Briefing series, 1973, available at https://soundcloud.com/user-262473248/a-sixty-minute -interview-with-leon-festinger.

"Suppose an individual...will happen?" and **"When the... frozen and expressionless":** Leon Festinger, Henry W. Riecken, and Stanley W. Schachter, *When Prophecy Fails: A Social and Psychological Study of a Modern Group That Predicted the Destruction of the World* (Minneapolis: University of Minnesota Press, 1956), 3, 162–63.

"I need not...points of the war": Charles Griffith, *The Quest: Haywood Hansell and American Strategic Bombing in*

World War II (Montgomery, AL: Air University Press, 1999), 132.

"The attacks on the ball-bearing...our last gasp": Albert Speer, *Inside the Third Reich: Memoirs by Albert Speer* (New York: Simon and Schuster, 1997), 286.

CHAPTER SIX: "IT WOULD BE SUICIDE, BOYS, SUICIDE."

"Our main job...terrible sometimes": Melvin S. Dalton, interview by Chris Simon for the Veterans History Project, American Folklife Center, Library of Congress, June 11, 2003, available at https://memory.loc.gov/diglib/vhp/story/loc.natlib.afc2001001.33401/sr0001001.stream.

"It was a lot of rocks...hot everywhere": Vivian Slawinski, interview by Jerri Donohue, Veterans History Project, American Folklife Center, Library of Congress, n.d., available at https://memory.loc.gov/diglib/vhp/story/loc.natlib.afc2001001.46299/sr0001001.stream.

"The beach here...had in Hawaii": Letter from Curtis LeMay to Helen LeMay, February 5, 1945, in Benjamin Paul Hegi, *From Wright Field, Ohio, to Hokkaido, Japan: General Curtis E. LeMay's Letters to His Wife Helen, 1941–1945* (Denton, TX: University of North Texas Press, 2015), 319.

Haywood Hansell's quotations and the cadet's quotation in this chapter, unless otherwise noted, are from Haywood Hansell, talk at the United States Air Force Academy, April

19, 1967, Clark Special Collections Branch, McDermott Library, United States Air Force Academy, Colorado Springs, CO.

"Stick together… *on the target*": Charles Griffith, *The Quest: Haywood Hansell and American Strategic Bombing in World War II* (Montgomery, AL: Air University Press, 1999), 175.

"It was a grueling hell…their flight path": Curtis LeMay and Bill Yenne, *Superfortress: The Boeing B-29 and American Air Power in World War II* (New York: McGraw-Hill, 1988), 72.

"That was a crazy thing…insane" and **"When they started flying…just exhausted":** David Braden, interview by Alfred F. Hurley, Dallas, TX, February 4, 2005, University of North Texas Library, Denton, TX, available at https://digital.library.unt.edu/ark:/67531/metadc306702/?q=david%20braden.

"Listen to me, boys…suicide": 40th Bomb Group Association, "An Ersatz Tokyo Rose Intro," available at http://40thbombgroup.org/sound2.html.

"I'd rather have somebody…do nothing": Reminiscences of Curtis E. LeMay: Oral History, 1971 (Air Force Academy Project, Columbia Center for Oral History, Columbia University Libraries, New York, NY).

Information about San Antonio One and other bombing missions can be found in *The Army Air Forces in World War II*, ed. Wesley Frank Craven and James Lea Cate, vol. 5, *The Pacific: Matterhorn to Nagasaki, June 1944 to August 1945* (Washington, DC: Office of Air Force History, 1983),

557, available at https://media.defense.gov/2010/Nov/05/
2001329890/-1/-1/0/AFD-101105-012.pdf; and Harry A.
Stewart, John E. Power, and United States Army Air
Forces, "The Long Haul: The Story of the 497th Bomber
Group (VH)" (1947). World War Regimental Histories.
106. http://digicom.bpl.lib.me.us/ww_reg_his/106.

"Six hours later...waiting for?": William Keighley,
dir., *Target Tokyo* (Culver City, CA: Army Air
Forces First Motion Picture Unit, 1945), available at
https://www.pbs.org/wgbh/americanexperience/features/
pacific-target-tokyo/.

Lieutenant Ed Hiatt's quotations are from Elaine Donnelly
Pieper and John Groom, dirs., *The Jet Stream and Us*
(Glasgow: BBC Scotland, 2008).

Information about weather balloons comes from "Weather
Balloons," Birmingham, Alabama, Weather Forecast Office,
National Weather Service, available at https://www
.weather.gov/bmx/kidscorner_weatherballoons.

Information about the jet stream, Rossby waves, and
Wiley Post comes from "The Carl-Gustaf Rossby Research Medal," American Meteorological Society, available
at https://www.ametsoc.org/index.cfm/ams/about-ams/ams-a
wards-honors/awards/science-and-technology-medals/the-ca
rl-gustaf-rossby-research-medal/; "Post, Wiley Hardeman,"
National Aviation Hall of Fame, available at https://www.na
tionalaviation.org/our-enshrinees/post-wiley-hardeman/; and
Tom Skilling, "Ask Tom Why: Who Coined the Term Jet
Stream and When?," *Chicago Tribune,* September 23, 2011.

"And Jesus...tempted by the devil" and "And the devil...'it will all be yours'": Luke 4:1–2 and Luke 4:5–7, English Standard Version.

CHAPTER SEVEN: "IF YOU, THEN, WILL WORSHIP ME, IT WILL ALL BE YOURS."

Hoyt Hottel's quotations are from Hoyt Hottel, interview by James J. Bohning, Cambridge, MA, November–December 1985, Center for Oral History, Science History Institute, available at https://oh.sciencehistory.org/oral-histories/hottel-hoyt-c.

William von Eggers Doering's quotations are from William von Eggers Doering, interview by James J. Bohning, Philadelphia, PA, and Cambridge, MA, November 1990 and May 1991, Center for Oral History, Science History Institute, available at https://oh.sciencehistory.org/oral-histories/doering-william-von-eggers.

Louis Fieser's quotations are from Louis F. Fieser, *The Scientific Method: A Personal Account of Unusual Projects in War and in Peace* (New York: Reinhold, 1964).

For more information about the birth of napalm, see Robert M. Neer, *Napalm: An American Biography* (Cambridge, MA: Belknap Press, 2015).

"After some considerable...for London": Charles L. McNichols and Clayton D. Carus, "One Way to Cripple Japan: The Inflammable Cities of Osaka Bay," *Harper's Magazine* 185, no. 1105 (June 1942): 33.

For more information about the tests at Dugway, see Standard Oil Development Company, "Design and Construction of Typical German and Japanese Test Structures at Dugway Proving Ground, Utah" (1943), available at https://drive.google.com/file/d/1eiqYwvJNS Y-ZpUsNQozwBISyQv_z4Uzb/view.

For the NDRC's analysis of incendiary weapons, see National Defense Research Committee, *Summary Technical Report of Division 11*, vol. 3, *Fire Warfare: Incendiaries and Flame Throwers* (Washington, DC, 1946), available at https://www.japanairraids.org/?page_id=1095.

"The main component...toward the target": *M-69 Incendiary Bomb*, Department of Defense combat bulletin no. 48, PIN 20311, 1945, available at https://www.youtube.com/watch?v=uPteVZyF4U0.

"Boys. It's tough...doing well" and "I don't agree... operation will fail": Transcript of Interview with Major General J. B. Montgomery, Los Angeles, CA, August 8, 1974, Clark Special Collections Branch, McDermott Library, US Air Force Academy, Colorado Springs, CO.

"an urgent requirement for planning purposes": Charles Griffith, *The Quest: Haywood Hansell and American Strategic Bombing in World War II* (Montgomery, AL: Air University Press, 1999), 182.

"Every important building...the next day": William W. Ralph, "Improvised Destruction: Arnold, LeMay, and the Firebombing of Japan," *War in History* 13, no. 4 (2006): 517, doi:10.1177/0968344506069971.

CHAPTER EIGHT: "IT'S ALL ASHES. ALL THAT
AND THAT AND THAT."

Many of the primary sources cited in this chapter and elsewhere are available at Japan Air Raids (https://www.japanairraids.org/), a bilingual historical archive run by David Fedman, assistant professor of East Asian history at the University of California at Irvine, and Cary Karacas.

"How many times…we'll go" and "Suddenly, in the air…piercing the air": Curtis E. LeMay with MacKinlay Kantor, *Mission with LeMay: My Story* (New York: Doubleday, 1965), 13–14, 351.

"And there was just a gasp…low-altitude [flight profile]" and "Frankly, when those…fire that big": David Braden, interview by Alfred F. Hurley, Dallas, TX, February 4, 2005, University of North Texas Library, Denton, TX, available at https://digital.library.unt.edu/ark:/67531/metadc306702/?q=david%20braden.

"I have been asked…personal decision": Haywood Hansell, talk at the United States Air Force Academy, April 19, 1967, Clark Special Collections Branch, McDermott Library, United States Air Force Academy, Colorado Springs, CO.

Curtis LeMay's quotations in this chapter, unless otherwise noted, are from Reminiscences of Curtis E. LeMay: Oral History, 1971 (Air Force Academy Project, Columbia Center for Oral History, Columbia University Libraries, New York, NY).

"Colonel LeMay...very inaccurate": *First U.S. Raid on Germany,* Reuters, British Pathé newsreel, 1943, available at https://www.youtube.com/watch?v=YgO6DX_9z0I.

"It worked out...but it worked" and "War is a mean...quick as possible": Curtis LeMay oral history interview, March 1965, Air Force Historical Research Agency, Montgomery, AL.

"Ralph, you're probably...much better": Emily Newburger, "Call to Arms," *Harvard Law Today,* October 1, 2001, available at https://today.law.harvard.edu/feature/call-arms/.

St. Clair McKelway's quotations are from St. Clair McKelway, "A Reporter with the B-29s: III—The Cigar, the Three Wings, and the Low-Level Attacks," *The New Yorker,* June 23, 1945, 26–39.

"We had a good mission...grocery bill for a month": Letter from Curtis LeMay to Helen LeMay, March 12, 1945, in Benjamin Paul Hegi, *From Wright Field, Ohio, to Hokkaido, Japan: General Curtis E. LeMay's Letters to His Wife Helen, 1941–1945* (Denton, TX: University of North Texas Libraries, 2015), 330.

"There were a couple...that kind of thing": Jane LeMay Lodge, interview by Barbara W. Sommer, San Juan Capistrano, CA, September 10, 1998, Nebraska State Historical Society, available at http://d1vmz9r13e2j4x.cloudfront.net/nebstudies/0904_0302jane.pdf.

"the temptation...greatness of the end": George Slatyer Barrett, *The Temptation of Christ* (Edinburgh: Macniven & Wallace, 1883), 48.

Information about the impact of the March 10, 1945, Tokyo bombing is from R. Cargill Hall, ed., *Case Studies in Strategic Bombardment* (Washington, DC: Air Force History and Museums Program, 1998), 319, available at https://media.defense.gov/2010/Oct/12/2001330115/-1/-1/0/AFD-101012-036.pdf.

"That the more densely … 'paper and ply-board structures'": David Fedman, "Mapping Armageddon: The Cartography of Ruin in Occupied Japan," *The Portolan* 92 (Spring 2015): 16.

"Probably more…history of man": United States Strategic Bombing Survey, *A Report on Physical Damage in Japan,* June 1947, 95, available at https://dl.ndl.go.jp/info:ndljp/pid/8822320.

CHAPTER NINE: "IMPROVISED DESTRUCTION."

Information about LeMay's bombing of Japan in the spring of 1945 is from C. Peter Chen, "Bombing of Tokyo and Other Cities: 19 Feb 1945–10 Aug 1945," World War II Database, available at https://ww2db.com/battle_spec.php?battle_id=217.

"Our B-29s…Japan had capitulated": Curtis E. LeMay with MacKinlay Kantor, *Mission with LeMay: My Story* (New York: Doubleday, 1965), 388.

Curtis LeMay's quotations in this chapter, unless otherwise noted, are from Reminiscences of Curtis E. LeMay: Oral

History, 1971 (Air Force Academy Project, Columbia Center for Oral History, Columbia University Libraries, New York, NY).

"What a kick…feeling fine": J. W. Stilwell diary, September 1, 1945, quoted in Jon Thares Davidann, *The Limits of Westernization: American and East Asian Intellectuals Create Modernity, 1860–1960* (New York: Taylor & Francis, 2019), 208.

"Was it possible…enemy civilians?": Ronald Schaffer, *Wings of Judgment: American Bombing in World War II* (Oxford, UK: Oxford University Press, 1985), 180.

"Deliberate terror bombing…hasten Hitler's doom" and **"Our policy…civilian populations":** Mark Selden, "A Forgotten Holocaust: US Bombing Strategy, the Destruction of Japanese Cities, and the American Way of War from World War II to Iraq," *Asia-Pacific Journal: Japan Focus* 5, no. 5 (May 2, 2007), available at https://apjjf.org/-Mark-Selden/2414/article.html.

"We have discovered…old capital or the new": Erik Slavin, "When the President Said Yes to the Bomb: Truman's Diaries Reveal No Hesitation, Some Regret," *Stars and Stripes,* August 5, 2015.

"It is striking…from above?": William W. Ralph, "Improvised Destruction: Arnold, LeMay, and the Firebombing of Japan," *War in History* 13, no. 4 (2006): 517, doi:10.1177/0968344506069971.

"Bygones are bygones…our Air Self-Defense Units": Robert Trumbull, "Honor to LeMay by Japan Stirs Parliament

Debate," *New York Times,* December 8, 1964, available at https://timesmachine.nytimes.com/timesmachine/1964/12/08/99401959.html?pageNumber=15.

Information about the dispute between George C. Marshall and Ernest J. King is from Richard B. Frank, "No Recipe for Victory," National WWII Museum, August 3, 2020, available at https://www.nationalww2museum.org/war/articles/victory-in-japan-army-navy-1945.

"In his autobiography…something he could have said": Warren Kozak, *LeMay: The Life and Wars of General Curtis LeMay* (Washington, DC: Regnery Publishing, 2009), 341.

CONCLUSION: "ALL OF A SUDDEN, THE AIR HOUSE WOULD BE GONE. POOF."

Information about the United Nations protocol banning incendiary weapons is from "Protocol III to the Convention on Prohibitions or Restrictions on the Use of Certain Conventional Weapons Which May Be Deemed to Be Excessively Injurious or to Have Indiscriminate Effects," United Nations Office for Disarmament Affairs Treaties Database, available at http://disarmament.un.org/treaties/t/ccwc_p3/text.

Index

Also by Malcolm Gladwell

Talking to Strangers

The routine traffic stop that ends in tragedy. The spy who spends years undetected at the highest levels of the Pentagon. The false conviction of Amanda Knox. Why do we so often get other people wrong? Why is it so hard to detect a lie, read a face or judge a stranger's motives?

Using stories of deceit and fatal errors to cast doubt on our strategies for dealing with the unknown, Malcolm Gladwell takes us on an intellectual adventure into the darker side of human nature, where strangers are never simple and misreading them can have disastrous consequences.

'Compelling, haunting, tragic stories . . . resonate long after you put the book down' James McConnachie, *The Sunday Times*

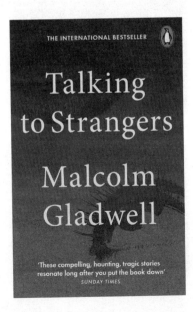

Also by Malcolm Gladwell

David and Goliath

What if everything we thought about power was wrong? *David and Goliath* is about what happens when ordinary people confront giants. Malcolm Gladwell weaves unforgettable stories of misfits, outsiders, tricksters and underdogs who have faced outsize challenges and won. With his gift for showing us the world through new eyes, he lets us see why the powerful are not as powerful as they seem, and that some of us have more strength and purpose than we could imagine. It is the story of how overwhelming odds can produce greatness and beauty.

'Gladwell is a master craftsman, an outlier amongst authors'
Huffington Post

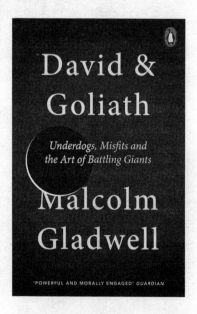

Also by Malcolm Gladwell

What The Dog Saw

In *What The Dog Saw* Malcolm Gladwell shows how the most ordinary subjects can illuminate the most extraordinary things about us and our world. Looking under the surface of the seemingly mundane, he explores the underdogs, the overlooked, the curious, the miraculous and the disastrous, and reveals how everyone and everything contains an incredible story.

'He is the best kind of writer – the kind who makes you feel like you're a genius, rather than he's a genius' *The Times*

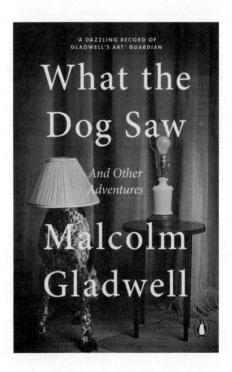

Also by Malcolm Gladwell

Outliers

Why do some people achieve so much more than others? Can they lie so far out of the ordinary? *Outliers* reveals that the story of success is far more surprising, and more fascinating, than we could ever have imagined. It will change the way you think about your own life story, and about what makes us all unique.

'Gladwell makes the world seem fresh and exciting again' *Evening Standard*